Eastern Washington's

Vanished Gristmills

And the Men Who Ran Them

Florence E. Sherfey

Eastern Washington's

VANISHED GRISTMILLS

AND THE MEN WHO RAN THEM

FLORENCE E. SHERFEY

YE GALLEON PRESS
FAIRFIELD, WASHINGTON
1978

Library of Congress Cataloging in Publication Data

Sherfey, Florence E

Eastern Washington's vanished gristmills and the men who ran them.

Bibliography: p.

Includes index.

1. Flour-mills--Washington (State) --History. 2. Washington (State)--Industries. I. Title. II. Title: Vanished gristmills and the men who ran them.

TS2135.U6W27 664'.72 78-21606

ISBN 0-87770-208-x

DEDICATED TO:

the Washington Territory pioneers who long ago operated these water-powered gristmills......

Vanished Gristmills

STATE OF WASHINGTON

Eastern Counties with Gristmills

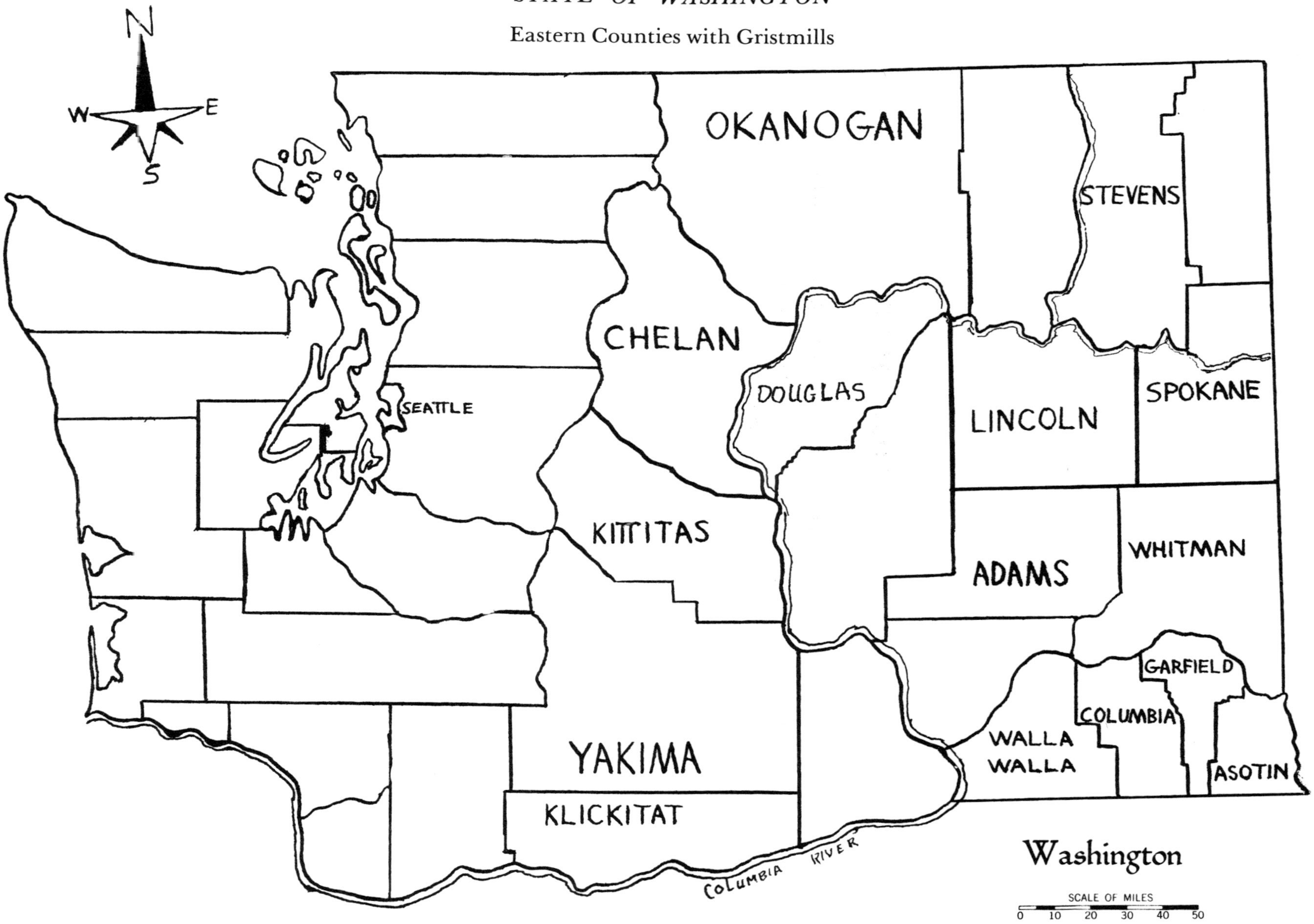

Vanished Gristmills

EASTERN WASHINGTON GRISTMILLS

IN CHRONOLOGICAL ORDER OF EXISTENCE

Built before 1900

Date Built	*Name of Gristmill*	*County*
1816	NW Fur Co. Gristmill at Marcus Flats	Stevens
1838	Marcus Whitman (tubmill)	Walla Walla
1840	Spalding (tubmill) on Alpowa; later Silcott	Asotin
1843	HBCo. Gristmill, Goudy Mill, Meyers Falls Mill	Stevens
1859	Oppenheimer Gristmill (by Judge Yantis)	Stevens
1859	Pioneer Gristmill	Walla Walla
1860	Douglas Gristmill (Operated short time)	Stevens
1862	Frontier	Walla Walla
1862	Coyle's (originally Frontier)	Walla Walla
1862	North Pacific Flour Mills	Walla Walla
1865	Wait's Mill, Waitsburg	Walla Walla
1866	Long's Mill	Walla Walla
1866	Reese's Standard Flour Mill (Later McCalley)	Walla Walla
1868	Chewelah Flour Mill	Stevens
1869	Schanno Mill at Union Gap	Yakima
1871	Adams Gristmill	Stevens
1871	Dayton Flour Mills (Wait's)	Columbia
1872	Meyers Falls Flour Mill	Stevens
1872	Phoenix Flour Mill	Walla Walla
1872	Simcoe Mill	Yakima
1873	Colfax Flour Mill	Whitman
1875	Palouse Flour Mills	Whitman
1875	Brick Mill, near Ellensburg	Kittitas
1877	Almota Flour Mill	Whitman
1877	Chase's Mill at Lincoln	Whitman

Florence E. Sherfey

1877	Marengo	Columbia
1877	Columbia Center (settlement Garfield Co.)	Garfield
1877	Thorp Mill	Kittitas
1878	Pomeroy	Garfield
1878	Klickitat Mill	Klickitat
1879	Tjossem Mill	Kittitas
1879	Post's Flour Mill (Frederick Post)	Spokane
1879	Damman Mill near Ellensburg	Kittitas
1879	Houser Flour Mill (Later Pataha City Fl. Mills)	Garfield
187-	Eureka Mill (Exact date unknown)	Okanogan
187-	North Star (Exact date unknown,Twisp)	Okanogan
187-	Addy	Stevens
1880	Wagner & Dybdall, later Dybdall	Spokane
1880	Deep Creek Gristmill	Spokane
1880	Alpowa Gristmill	Garfield
1880	Albion	Whitman
1881	Asotin Flour Mills, later Jerry Milling Co.	Asotin
1881	Pine City Milling Co.	Whitman
1881	Echo Flour Mills	Spokane
1882	Colton Flour Mill	Whitman
1882	Corbett Mill at Huntsville	Walla Walla
1882	Silcott (Woodworth Mill)	Asotin
1882	Brooklyn (settlement near Dayton)	Columbia
1883	Dartford Flour Mill on Little Spokane	Spokane
1883	Prescott Flour Mill	Walla Walla
1884	Elberton Flour Mill	Whitman
1884	Sprague Mill at Lincoln	Lincoln
1885	C & C Flour Mill	Spokane
1887	Marshall Milling co.	Spokane
1887	City Mills at Ellensburg	Kittitas
1887	Nashland in Big Bend	-----?
1888	Gould City (settlement near Pomeroy)	Garfield
1888	Sanders Flour Mill	Kittitas
188-	Winona (date unknown)	Whitman
188-	Albion	Whitman
1888	Garfield Flour Mill	Whitman
1888	Thornton Flour Mill	Whitman

Vanished Gristmills

1889	Holmes Spur Mill	Kittitas
1889	Centennial Flour Mills	Spokane
1889	Rockford Flour Mill	Spokane
1889	Reardan Flour Mill	Lincoln
1890	Joe Henry's Flour Mill (Joseph Heinrich)	Spokane
1890	Cheney Flour Mill	Spokane
1890	Pullman Flour Mill	Whitman
1890	Oakesdale Mill	Whitman
1892	Colfax Milling Co.	Whitman
1898	City Mills	Walla Walla
1898	Acme Mill at Lamar	Walla Walla
189-	Tekoa (exact date unknown)	Whitman
1890	Wilbur	Lincoln
1890	Anatone Flour Mill	Asotin
1893	Wenatchee Roller Mills	Wenatchee
1891	Waterville Milling Co.	Chelan
1897	Ritzville Flour Mills	Lincoln
1899	Harrington Milling Co.	Lincoln
189-	Union Town Mill	Whitman

Florence E. Sherfey

GRISTMILLS BY COUNTIES

STEVENS COUNTY

Northwest Fur Company Mill
Hudson's Bay Mill at Ft. Colville
Meyers Falls Gristmill
Oppenheimer Mill
Douglass Mill
Addy Mill
Chewelah Flour Mill
Adams Gristmill

WALLA WALLA COUNTY

Pioneer Gristmill
Frontier Gristmill
Coyle's Mill
Wait's Mill
North Pacific Mills
Prescott Flour Mill
City Mills
McCalley, or Reese's Standard
Acme Mill at Lamar
Covello
Phoenix

ASOTIN COUNTY

Silcott (originally Alpowa City)
Asotin Flour Mills, (Jerry Milling)
Anatonc Flour Mills
Jerry Mill at Jerry

SPOKANE COUNTY

Post's Gristmill in Spokane
C & C Flour Mill
Echo Flour Mills
Centennial Flour Mills
Dybdall Mill, Chapman Lake, Cheney
Dartford Gristmill, Little Spokane
Deep Creek (J.A. Bauman and Sons)
Marshall Milling Company
Joe Henry's Gristmill near Fairfield
Cheney Flour Mill
Cox's Mill
Rockford Flour Mill

GARFIELD COUNTY

Columbia Center Gristmill
Pomeroy Flour Mill
Pataha City Flour Mills
Alpowa Gristmill
Gould City Mill

COLUMBIA COUNTY

Dayton Flour Mill (Wait's)
Marengo Gristmill
Corbett Mill at Huntsville
Brooklyn Mill
Long's Mill

Vanished Gristmills

WHITMAN COUNTY

Palouse Flour Mills
Colfax Flour Mills
Pine City Milling Co.
Almota Flour Mill
Chase's Mill at Lincoln
Elberton Flour Mill
Oakesdale (J.C. Barron Flour Mill)
Garfield Flour Mill
Albion Mill
Winona Flour Mill
Farmington Flour Mill
Union Town Mill
Tekoa Flour Mill
Pullman Flour Mill
Rosalia Mill
LaCrosse Flour Mill
St. John Flour Mill
Colton Gristmill
Thornton Gristmill

OKANOGAN COUNTY

North Star Mill at Twisp

KLICKITAT COUNTY

Klickitat Mill at Goldendale

CHELAN COUNTY

Wenatchee Flour Mill

DOUGLAS COUNTY

Waterville Milling Co.

KITTITAS COUNTY

Tjossem Mill near Ellensburg
Sanders Gristmill
City Flour Mill
Holmes Spur Mill
Brick Mill
Damman Flour Mill
Thorp Gristmill

LINCOLN COUNTY

Sprague Gristmill at Lincoln
Reardan Flour Mills
Wilbur Flour Mill
Harrington Flour Mill

ADAMS COUNTY

Ritzville Flour Mill

YAKIMA COUNTY

Union Gap, Schanno Flour Mill
Simcoe Mill, Yakima Reserv.

MISCELLANEOUS

Wallowa County in Oregon
Troy Flour Mill

Post Falls, Idaho
Cable Mill

Vancouver, Wash.
Cedar Gap Mill
on Cedar Creek

Florence E. Sherfey

CONTENTS

Maps by John F. Carter

Endsheets by author, from a photograph submitted by Mrs. Wyoma Duckworth.

Unless otherwise stated, illustrations and photographs supplied by author.

Vanished Gristmills

Introduction

The gristmills included in this volume are the early water-powered gristmills in Washington, mainly east of the Cascade Mountains and built before 1900. Very little visual documentation exists of these early gristmills, since only a scant handful of them still stand.

In trying to find information about them it was evident from the beginning of my search, that I was about three generations too late, for those who held the real facts and knowledge about the old mills were gone.

A few oldsters still living remember about some of them. People ninety-four and ninety-seven gave valuable information, with one telling of her grandfather, who was a builder of gristmills. Another said, "My father carried grist to the mill on the creek," and one recalled, "That old mill standing is where we got our flour; it wasn't always the lightest, but that's what we used." Few of these old-timers can actually describe the mill or give a visual picture of its interior mainly because they were too young then for those details to have registered with them. Besides, everyone took a gristmill for granted. The family carried the wheat there and brought back the whole wheat flour, the miller having run the grains of wheat down into the stones where they ground slowly at the rate of about a bushel to a half hour.

Many pictures of the old gristmills have come to me. A number of people have sent me useful information, or have given me details of the old mills, all for which I am extremely grateful. Gristmills are of interest to nearly everyone, for undoubtedly all of our great grandparents, as well as others of our early ancestors ate bread and other baked goods made from flour ground out by these mills. Besides that, they hold an interest all their own.

Each mill was different in that it had its own individuality, being built to suit the location of the area and the farmers' needs, but with their inner workings, their equipment and machinery all basically of the same plan. The exterior of the old buildings all show a similar huge bulkiness that marks each one unmistakably a gristmill or an old flour mill. They were built strictly on a utilitarian basis without the slightest pretence to eye appeal, and yet I still have to see a picture of one, or any mill standing, which is not beautiful. They rest boldly and proudly against the skylines wearing the raiments of their broken shabbiness as proudly as a royal courtier displaying his rich garments. The old mills' thick wooden sidings seldom knew the luxury of a coat of handmade paint, forming their own stains instead, of colors from nature's muted palette—soft greys, warm tones of ochre and earth reds and browns, the oils of the woods combining with the natural pigments from wind, dust and weather.

None of the old water wheels seems to have survived. I have yet to find a picture of an Eastern Washington gristmill showing the old water wheel, with the exception of the picture of Joe Henry's gristmill near Fairfield. A small section of his overshot water wheel shows slightly at the left back of the mill. Most of the water wheels, after a few years—ten or fifteen, possibly, were removed and the power 'stepped up' with the addition of a turbine. The water wheel quickly decomposed from its exposure to moisture and air.

Some of the gristmills were steam-powered, especially if a strong stream was not available to supply water power. Use of steam ever since its invention by James Watt in 1769, was a close competitor of water for early gristmills. By 1890, steam was used in 5,383 of the 22,573 gristmills in America, often as a standby source of power.

It will unquestionably sadden me when I discover some gristmills may not have been included in this volume. I would have wanted them here if possible. It would be easy to miss some. Gristmills were extremely plentiful in those early times and built in nearly every little community, especially in southeastern Washington. People stayed close to their homes in those days, some never leaving their community, and a gristmill was a necessity since traveling was a slow, tedious process with horse and wagon.

The slow process of traveling was quickly altered when the railroads began grading track beds for a mechanical replacement of the faithful horse to carry people, and wheat. The ease of travel must have astonished country folks who could not have anticipated this in its fullest, probably not even realizing the change it would make in their lives, in their towns, and in the use and outcome of their gristmills.

When the gristmill's operation was at its height, the town was also flourishing, and when the towns began to dwindle, the gristmills were the first to go,lagging for awhile and then stopping altogether.

People literally walked away from them without a backward look, glad to be rid of the slow, cumbersome stones and their clouds of flour dust, for now they were taking their wheat to the larger mills in bigger towns. They discarded the old gristmills as they discarded their old sadirons, the washboard, the old wood ranges, the pot-bellied "heater" (which pa put up in the fall and took down in the summer). It went the way of the lantern, the horse and buggy, the old sod turner and many old tools which suddenly were useless.

Introduction

Like all those long-ago necessities, —now cherished in this generation—the gristmill is considered to be in the antiquity class; choice, studied, with the few standing placed in the "Keep for Posterity" category.

And they should be preserved, for gristmills represented an important part of peoples' world and lives. They fathered the first industry which has become the vast, complex one of flour milling.

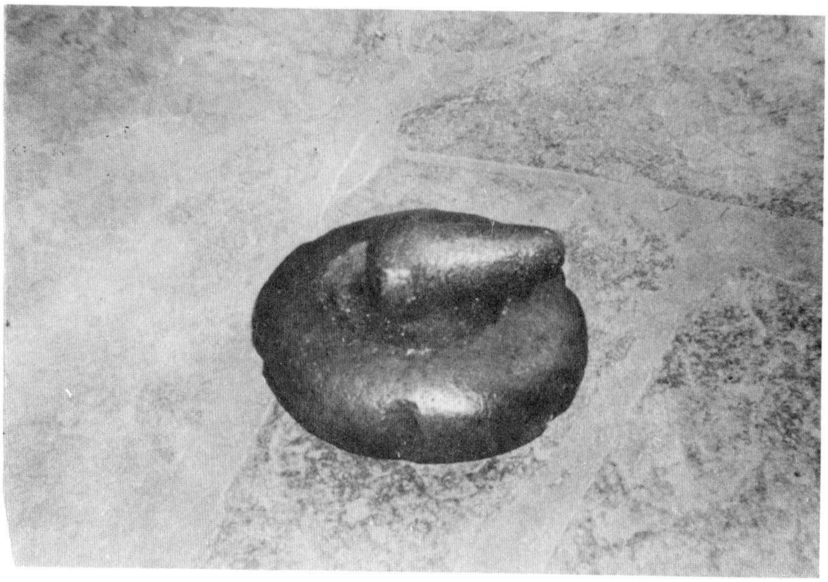

Mortar and pestle.
Courtesy of Mr. and Mrs. George T. Dart.

These grinding stones and several stones for pounding found along Columbia River, and used by Indians. The George Darts at Dartford have collected these.

How Gristmills Developed

THE growing of wheat and the milling of grain is almost as ancient as civilized man, for wheat has been in existence since about ten or fifteen thousand years before Christ.

Man early discovered that his wheat tasted better and was easier to chew when pounded and mashed. Perhaps at some time a little of this became moistened and was accidently left close enough to the fire to become baked into a sort of cake. From then on, the early milling of wheat into a meal and flour would have become a common practice among primitive cultures.

Stones were used for grinding by hand at first, but later the gristmills evolved, employing mechanical means, mainly water, to turn the mill stones for man.

The first millstone was any suitable stone with another stone as a pounder, to mash the grain. Some of these early stones were called *saddlestone*, the *mortar and pestle* and the *quern*.

The quern was the direct ancestor of the present millstone, but was not powered by water. Here two stones were placed one on top of the other. The top stone had a hole in its center into which grain was poured and another hole in its top into which a stick was inserted to turn it.

Other methods were devised to make the stones do their work with as little effort from man as possible. A tree became a *Sapling Mill* using the main branch of a young tree on which the stone was fastened. This could be pushed down into a basin of grain with the tree raising it up again. A hollowed out stump became a *Stump Mill* when grain was put in it and pounded with a stout branch end. A *Plump Mill* was a teeter-totter arrangement. Water from a trough filled a container at one end of a long piece of wood. When the end fell downward, heavy with the water, it raised the other end with the stone. As the water emptied out the stone plummeted downward into the basin of grain. When several of these were running in a neighborhood in early times, the plumping sounds made a great racket.

Much later, the nineteenth century farmer not only used his animals to plow his fields, but he invented ways in using them to mill his grain.

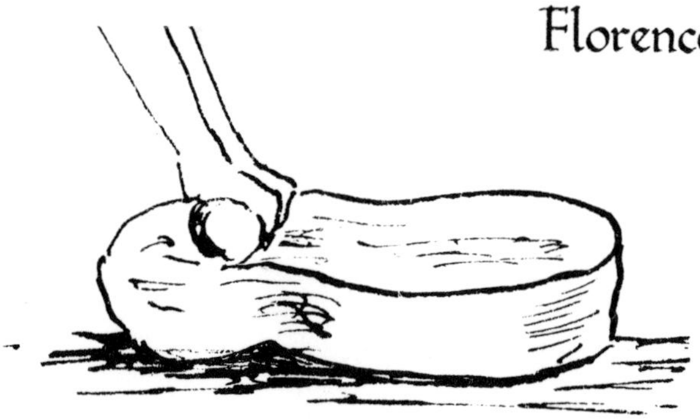

Saddlestone

Quern-type Hand Mill

Stump Mill

Sapling Mill

A *turntable* provided one method. It was a large wooden disk big enough for an animal to turn it by walking endlessly on it. This turning activated a wheel underneath it which was connected to machinery.

A *treadmill* could use one horse or two, walking forever on a platform and getting nowhere. But this constantly moving platform turned a big wheel belted to other wheels which could run a millstone, and often a horse could be hitched to push or pull a wooden beam which was fitted into a socket of a millstone to turn it. When several horses were used in this manner to turn a large millstone, a *sweepmill* resulted. More than wheat was ground between these stones, for they could grind flax into linseed oil, tobacco, salt, plaster or cider among many items used on a farm.

How Gristmills Developed

Plumping Mill

Comparing today's processes for producing flour with those of long ago, people living then must surely have appreciated their daily bread much more than we do today. Then there was no easy way to make this bread, for even producing the wheat itself was complicated hard work. The processes of farming were all done by hand, from planting to harvesting, to flailing the wheat from the grain heads, to winnowing out the dirt and chaff, and finally to grinding the wheat into flour.

It is believed that the first water-powered gristmills were in existence about nineteen B.C., first described then by the Roman engineer and architect, V. Pollio. The word *grist* meant, "as much grain as is carried to the mill at one time."

Gristmills were common in the eastern part of the United States as early as the 1600s. George Washington and Thomas Jefferson both owned gristmills, Jefferson as an investment and Washington running it as a business along with his farming. George Washington's mill is restored under the division of parks and operating today in Alexandria, Virginia. In Washington's diary he often noted, "Rid to mill today," or "Rid to Dogue Run today to look at mill dam at head of race." *

***Commonwealth of Virginia, Division of Parks.**

Florence E. Sherfey

In early 1800, animal power was used in areas not near a gristmill. An animal pushed or pulled a shaft attached to a millstone. When several horses pulled a large millstone, this was called a *sweepmill*.

One or two animals could walk on the moving platform of a *treadmill*. This would turn the large wheel belted to a smaller wheel, with its shaft fitted to machinery to run it.

A poor animal could run endlessly on a *turntable* which energized a wheel to turn a millstone, or run machinery.

How Gristmills Developed

When the first gristmills came to Eastern Washington, some as early as 1859 or before, they were all welcomed wholeheartedly by the farmers. By the 1870s they were being built in full force, and soon in the 1880s and 1890s, most every little settlement located on a stream boasted a water-powered gristmill.

Among the many new settlers to the West, mainly farmers looking for land, were men who came from the East trained in millwrighting and milling. A trained millwright had acquired special skills for building mills, and usually he had taken an apprenticeship in milling as well, both trades requiring knowledge and ability.

Word traveled quickly whenever there was a man in any region who was a millwright. He would usually be sought out and offered inducements to come to a settlement and build a gristmill. Farmers were more than willing to pledge their support to the mill, giving millsites, with housing for the miller included, and offer water rights.

A miller and millwright had several points to consider when choosing a location for a water-powered gristmill. First, the stream must have a fairly good drop from a higher to a lower level, and there must be a steady year round flow. He needed to know the quantity of water that flowed in a stream, and with this knowledge he would then determine the amount of power available. He must consider the irregularities in stream velocity or volume due to friction against the bed or banks, and must take into consideration also the climatic changes which would affect the flow. When these points were all calculated and understood, all it then took was plenty of stout lumber, a helper or two and a set, or run, of millstones, the granite ones usually ordered from France.

Water power was cheap and inexpensive power, and a gristmill was not beyond the financial means of most early towns, with everyone banding together to help in getting the mill underway.

Most all of the mill could be built of wood, with sometimes a stone basement. A sawmill or two usually operating nearby, furnished the lumber. The water wheel was made by the millwright, and, placed at the best location, provided the power to run all the equipment in the gristmill.

A water wheel was of three basic designs: *overshot, undershot,* and *breast wheel.* A *turbine* was a variation and probably the most efficient. Other variations were rose wheel, flutter wheels, tub wheels, float wheels, pitchback wheels, and others, but the water wheels used mainly in Eastern Washington gristmills were the overshot, undershot and the turbines.

The overshot wheel took water overhead from a flume, and was turned by the weight of the water, and usually used when a strong, steady and plentiful supply of water was available. The Encyclopedia Britannica states: "When a water fall ranges

Florence E. Sherfey

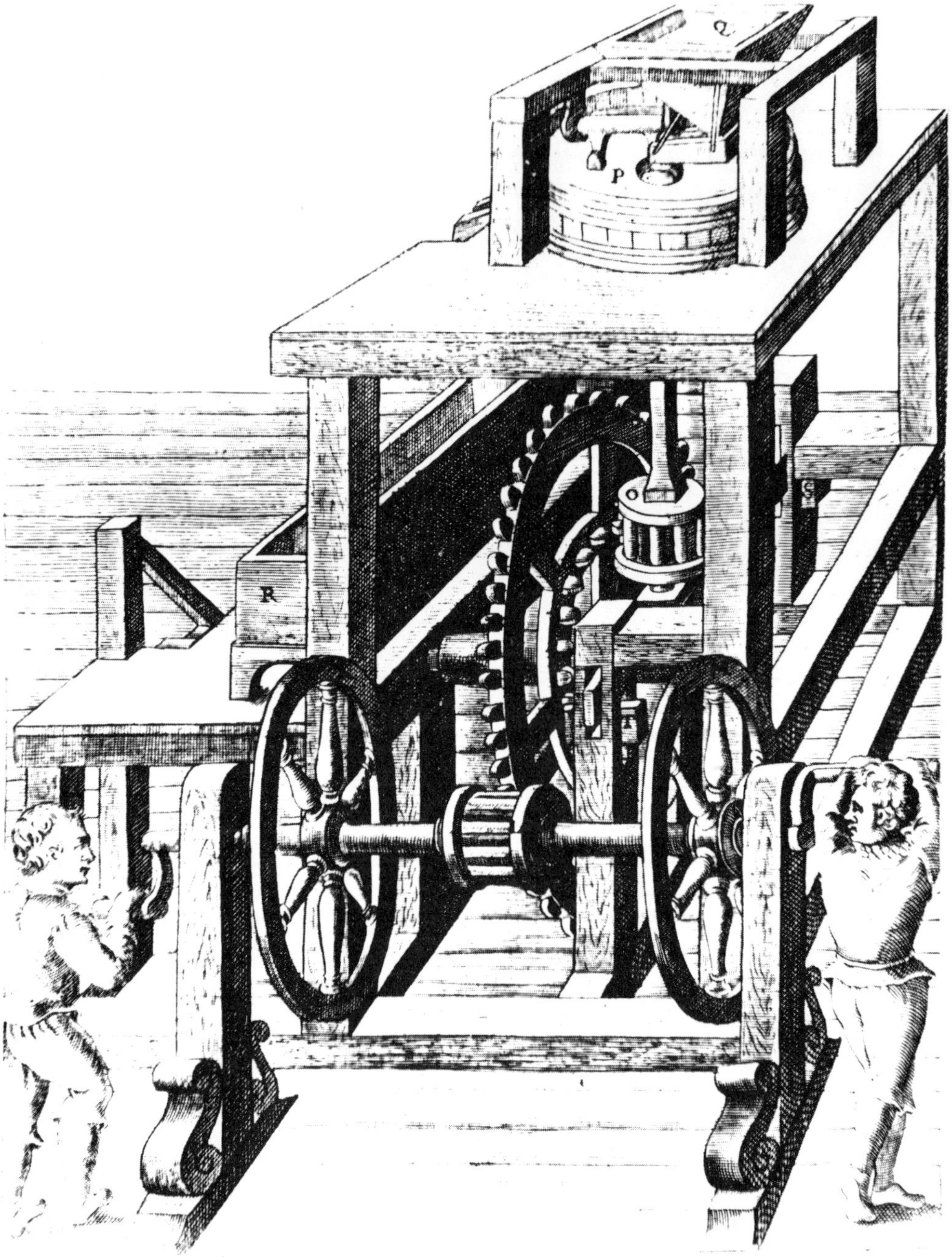

How Gristmills Developed

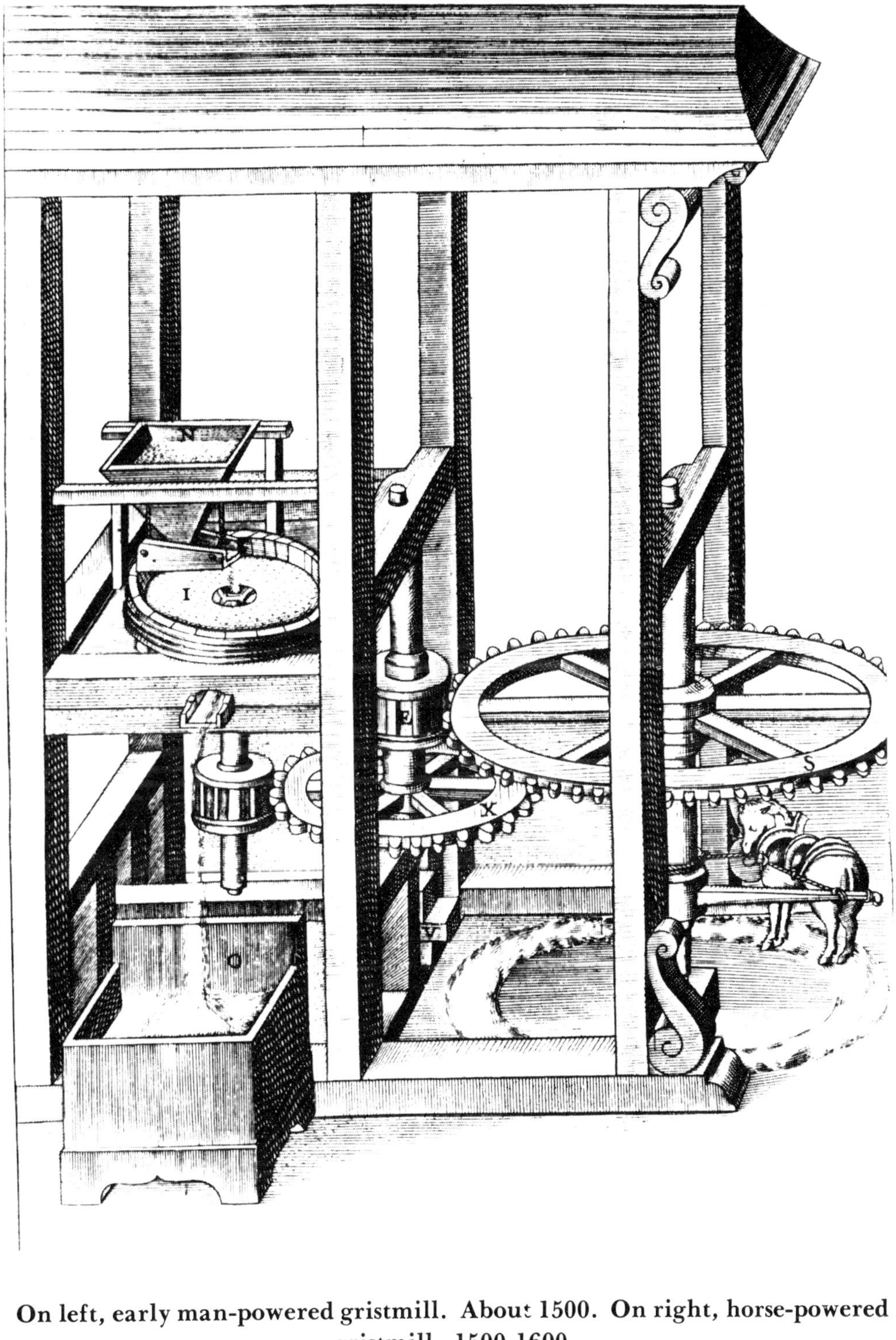

On left, early man-powered gristmill. About 1500. On right, horse-powered gristmill. 1500-1600.

(Courtesy Smithsonian Institution)

Florence E. Sherfey

Romans were credited with using the first wind and water mills. Water replaced wind for it was more reliable. No matter what the power, whether it was from windmills, animal power or any other power, the gearing came from a "master" wheel. The master wheel was connected to the main source of power by a shaft. In the windmills, the windsails were the source of power; in the water powered gristmill, either the water wheel, or the turbine created the power to run the mill.

between twenty and seventy feet and the water supply is from three to twenty-five cubic feet per second it is possible to construct a bucket wheel on which the water acts chiefly by its weight. If the variation of the head-water level does not exceed two feet, an overshot wheel may by used, with about twenty-four buckets to a ten-foot wheel." The overshot wheel was 75% efficient.

The undershot wheel was 35% efficient and took the water underneath, or at the bottom of the wheel, and usually used where the water was not a strong flow. Sometimes it took the water from a running stream but usually a dam was constructed near the stream and the water came to the wheel by way of a ditch and a flume. These wheels ordinarily had many paddles. Encyclopedia Britannica: "The diameter of the wheel should not be less than twice the fall, and is more often four times the fall."

A breast wheel received water at its hub or center. It was 65% efficient and carried a smaller load of water than the overshot wheel. George Washington's water wheel was a breast wheel. It was often used with the tide water at the ocean.

The turbine, a modification of the tub wheel, was the most efficient of all. This wheel was usually enclosed in a casing of cast iron, or wood, so that a great amount of pressure was created by the water within. The water came to the turbine down a vertical pipe called a penstock or standpipe. One of the advantages of the turbine, besides its greater power, was that it did not freeze in winter.

Soon after the turbine made its appearance, the old style wooden water wheel still continued in use, undoubtedly because it was cheaper to produce, being built by the millwright.

To keep the flow to the water wheels strong and provide a steady head, or fall, a dam (also called a millpond or reservoir) was nearly always built at the stream to collect and hold water. *Head* of water signified the distance the water dropped to reach the wheel, and the *fall* referred to the action of the water to turn the wheel. To carry the water from the dam towards the mill, a ditch was usually dug in the earth. The ditch was also called a millrace, millway, pipe, sometimes flume or penstock. Occasionally it was made of wood or concrete. When the water in the ditch reached a point near the gristmill, it was then taken over by another passageway to bring the water to the water wheel. This final passage was often constructed of wood and could have several names: wooden box flume, flume or pipe, millway, millrun, sluice, or sluiceway. It often had a sluice gate, which opened and closed to control the flow of water.

If the wooden box flume brought water to an overshot wheel, the flume was elevated, and if taking water to an undershot wheel, the flume might rest on the incline of a natural slope so the flow of water fell to the bottom of the wheel.

Overshot Water Wheel, 75 per cent efficient.
(Courtesy Smithsonian Institution)

How Gristmills Developed

Overshot Water Wheel

Cut-away of gristmill run by overshot water wheel. Wheel is about 20 feet in diameter, with drive wheel attached about 10 feet. (A) sluice gate, with man opening it. (B) drive or master wheel, transmitting power from Water Wheel. (E) Beveled gear with shaft carrying power to other gears and shafts to turn millstones and run other machinery.

(Courtesy Smithsonian Institution)

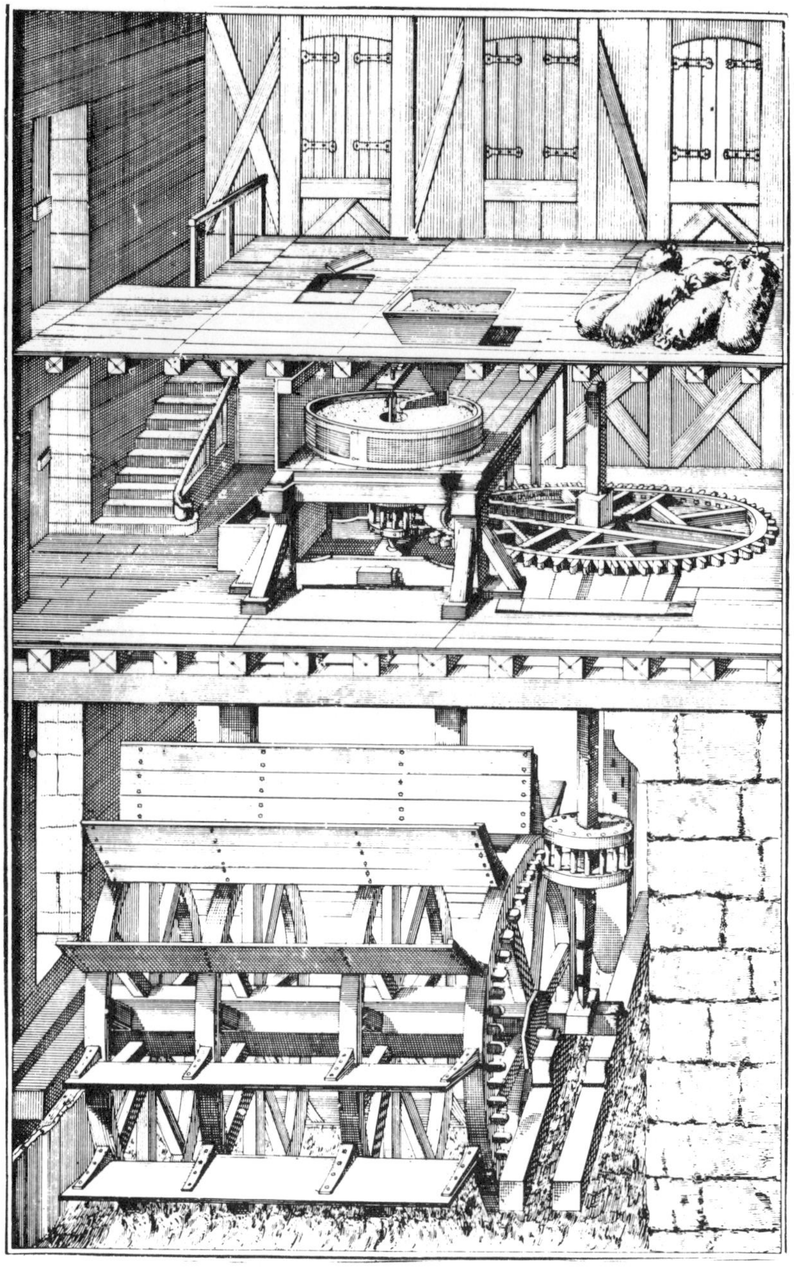

Early gristmill of the middle 1700s. Powered by undershot "paddle" wheel with gearing on face of water wheel. Lantern gear meshes with this, transmitting power to large gear above which energizes and runs millstones. Gravity played a large role in the workings of all mills. Wheat was taken to top floor, gradually "falling" to the selected operations until flour resulted.

(Courtesy Smithsonian Institution)

How Gristmills Developed

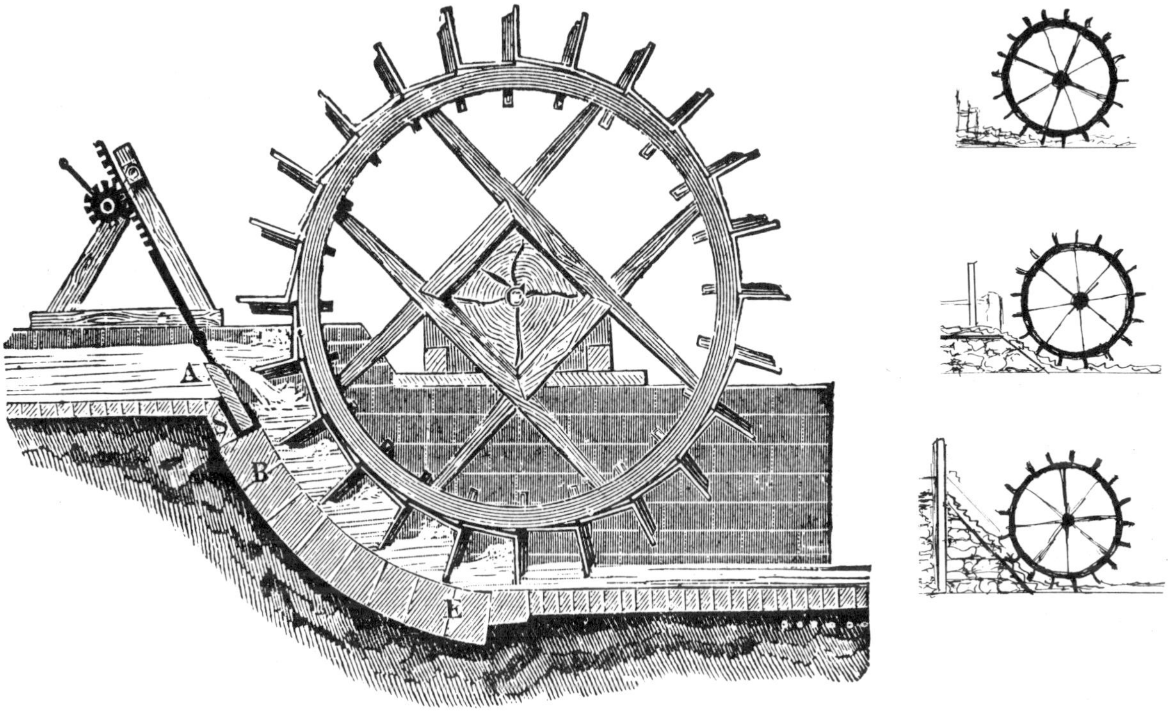

Undershot Water Wheel

Undershot wheels were sometimes used without a dam, with the fall of water adjusted to necessary height and angle at water wheel. In the illustration, (A) is the flume or sluice; (S) is sluice gate to adjust flow of water, and (B) and (E), the sluice way. Undershot wheels were 35 percent efficient.

(Photo, Smithsonian Institution).

Florence E. Sherfey

If the water was to go to a turbine, the final passageway (flume) shot the water down a vertical, or steep-angled *standpipe* connected to the turbine, whirling the paddles of the turbine's enclosed water wheel at great speed.

After the water was through with its work, it fell away usually entering the stream at another point, or collecting in a little pond. The used water was called the tailrace, backwash, spillway, no longer rushing but lazy as a well fed cat seeking a place to rest.

The power of the mill was transferred from the water wheel to the millstones in about the same way in the 1800s as it was in the 1600s. There were approximately three or four steps involved. A shaft, or axle from the water wheel entered the mill. This shaft, usually a large wooden one, had at its end a big wheel called the power wheel or master gear, which supplied the power for all the different parts of the mill. If the wheel had cogs, it meshed with the one or two other smaller wheels with cogs. If the masterwheel was belted, it powered a long shaft on which were fitted other belted wheels. These in turn revolved other shafts and wheels, one of them powering the millstones. Here, a shaft ending with a gear, locked into the eye of the top millstone and turned it.

Millstones always operated as a pair, or set, of two stones, their surfaces facing each other usually horizontally. This set, or pair of stones was called "a run." If there were two pairs, it was called "two-run." Granite millstones were often called burrstones, or the older term, buhrstones, and a gristmill using burrstones was often called a burrmill, or buhrmill.

Millstones had a language of their own. The *eye* was the center of the top millstone which was called the *runner*, for it was the one that turned. The bottom millstone, the *nether* stone, remained stationary. All millstones had grinding grooves on their surfaces radiating out from the center, called *furrows*. The space between the furrows was the *land*, showing the agricultural influence. Furrows must be kept sharp in order to cut the grain but not smash it, and the sharpening of the furrows was termed, *dressing* the stones. Either a miller dressed the stones himself or had a man, a *pecker* or *millstone pecker* do this. The pecker used a *millpick* for this work. It could either be an iron wedge, which he held with one hand tapping it with a hammer in the other hand, or a two-edged chisel grasped in one hand braced against the other wrist and steadied against a firm object to better control the chipping of the furrow's edge.

Most mills used a device to lift up the heavy stones. This was a set of grappling hooks which connected into two holes of the runner stone, moving it to convenient position so it could easily be worked on.

How Gristmills Developed

Undershot Water Wheel

Undershot Wheel, 35% efficient. Wheel turned by stream's speed. Usually used where flow of water not strong. Sometimes used with a fall of water.

Overshot Water Wheel

Overshot Wheel, 75% efficient. Turned by weight of water. Used where a strong supply of water existed

Overshot Water Wheel

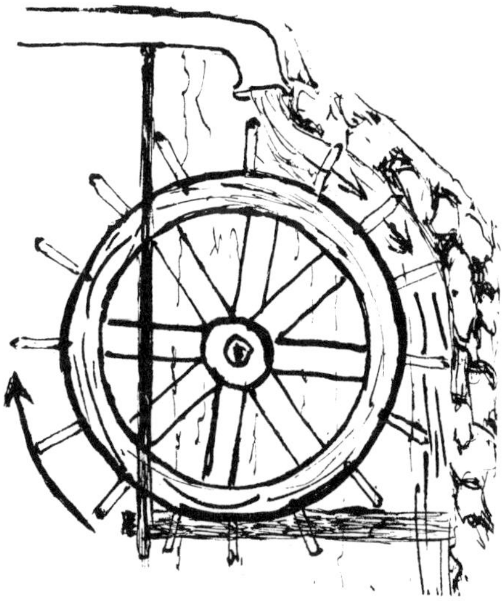

Breast Wheel, 65% efficient. Often used with tide, wheel turned by mass of water through a lock.

The stones, when grinding, were never permitted to touch each other for this would ruin the surfaces, and also the flour by generating too much heat. A miller had to watch constantly to see that the stones were separated, usually one-sixteenth of an inch, just far enough from each other for furrows to cut the grain and keep it moving. It is said a miller could always tell how far apart his stones were just by the sound.

Ventilation was important, otherwise a great deal of heat was generated which could cook the flour into a soggy product which could clog up the bolting silk, and also result in a poor loaf of bread. This heat contributed to the many fires in gristmills.

Dressing the stones

Grappling hooks for raising millstone for dressing the furrows.

How Gristmills Developed

Nether, or bottom millstone, cut in sections fitted together. Grooves, or furrows radiate out from center.

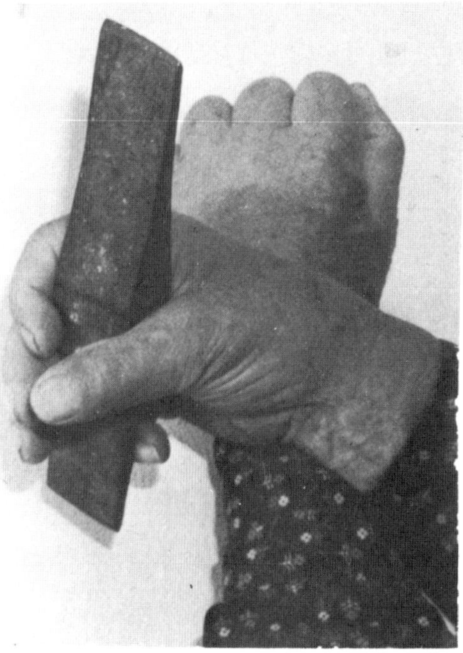

Millpick held in braced hands for better control in sharpening furrow edges of millstones.

Oftentimes, a miller tested the texture of his flour by feeling it, as a cook often used to feel her salt. He might also grasp a handful of flour and press it together. On opening his hand, if the flour flew out from his fingers he knew it "to be in a lively state," according to Oliver Evans, an early authority. If it were smooth and stuck to the hand it indicated that "the stones were set too low and the stones dull."

The early miller's rule-book was Oliver Evan's "The Young Millwright and Miller's Guide," published in 1795. Oliver Evans, born in Delaware in 1755, was a young genius who invented the automatic flour mill using the system of belt elevators and augers.

The surfaces of the millstones were slightly concave instead of flat, and the furrows were sharpened in the manner of scissors, the edges of one side opposing those of the other side. The grain was moved along in the furrows and worked gradually toward the outer edges, being ground finer and finer as it traveled to the feathered grooves at the rim, and finally thrown clear as flour.

TO FIND THE QUANTITY OF GRAIN WHEN HEAPED AGAINST A STRAIGHT WALL

Square one-half of the depth and proceed as in the previous rule.

INSIDE DIMENSIONS OF A FLOUR BARREL

Diameter at head, 17 inches. Diameter at bulge or middle, 20 inches. Depth, 26 inches. Volume of barrel, 4.15 cubic feet.

TO FIND THE QUANTITY OF GRAIN WHEN HEAPED ON THE FLOOR IN THE FORM OF A CONE

Square the depth and square the slant height, in inches. Take their difference and multiply by the depth, in inches. Multiply this product by .00048. The result will be the contents, in bushels.

TWO SIMPLE RULES FOR FINDING THE CAPACITY OF ROUND BINS, IN BUSHELS

Where the dimensions measure exactly in feet; multiply the diameter by the diameter, multiply by the depth, multiply by .63.

Where the dimensions measure in inches; multiply the diameter by the diameter, multiply by the depth, multiply by .000365.

CENTRIFUGAL FORCE FORMULAS

(In grams) = 4/980 × wt. in g × radius in cm. × π^2 × revs. per sec^2 = .0402836 × wt. in g. × radius in cm. × rev. per sec^2.

(In lbs.) = 4/32 × wt. in lbs. × radius in ft. × π^2 × rev. per sec^2 = 1.2337 × wt. in lbs. × radius in ft. × rev. per sec^2 = .102808 × wt. in lbs. × radius in inches × rev. per sec^2 = .0003410 × wt. in lbs. × radius in ft. × revs. per min.^{2}.

EAR CORN MEASURE

To find the contents of a corn crib, multiply the cubic feet by 4 and divide the product by 9*.

Find the contents of a corn crib, 18 feet long, 7 feet wide, 8 feet high.

*Note—This allows 2¼ cubic feet for a bushel. It is the rule most generally used, and will hold out in ordinary good corn, even if measured at the time it is cribbed.

7×8×18 = 1,008 cubic feet.
4

9)4032
Answer, 448 bushels.

CAPACITY OF BOXES, BINS, ETC.

By the U. S. standard, 2,150 cubic inches make a bushel; a cubic foot contains 1,728 cubic inches. Rule—Multiply the number of feet width of bin, by the length, the result by the depth, then divide the product by 5 and multiply the quotient by 4, which number will give the quantity in bushels.

Length	Breadth	Depth		Bushels
5′	3′	2′	will contain	24
5′	3′	3′	will contain	36
5′	3′	4′	will contain	48
7′	5′	3′ 9″	will contain	100
9′	6′	5′	will contain	216
[illegible]	8′	6′	will contain	500

...our feet eight inches long, two feet four inches ... feet four inches deep, will contain twenty

A box twenty-four inches by sixteen inches square, and twenty-eight inches deep, will contain a barrel.

A box twenty-six by fifteen and a half inches square, and eight inches deep, will contain a bushel.

A box twelve inches by eleven and a half inches square, and nine inches deep, will contain a half bushel.

A box eight by eight inches square, and eight inches deep, will contain a peck.

A box eight by eight inches square, and four and one-eighth inches deep, will contain one gallon.

A box seven by eight inches square, and four and one-eighth inches deep, will contain a half gallon.

A box four by four inches square, and four and one-fourth inches deep, will contain a quart.

To Compute the Contents of a Hopper—Multiply the length by the breadth, in inches, and this product by one-third the depth, measuring to the point. Divide the last product by 2,150 and the quotient thus obtained will be the contents of the hopper in bushels.

RUST PREVENTIVE

Parts of machines which are not in constant use and not actually painted should be coated with a rust preventive. A good and inexpensive coat may be made by mixing flour with common machinery or lubricating oil. This mixture should have the consistency of thin paste. It may be applied with a paint brush. This coating may be easily removed by a cloth or piece of waste saturated with coal oil or gasoline.

TABLE SHOWING U. S. STANDARD NUMBER OF THREADS TO AN INCH.

Diameter	¼	5/16	⅜	7/16	½	9/16	⅝	¾
No Threads	20	18	16	14	13	12	11	10
Diameter	⅞	1	1⅛	1¼	1⅜	1½	1⅝	1¾
No. Threads	9	8	7	7	6	6	5½	5
Diameter	1⅞	2	2¼	2½	2¾	3	3¼	3½
No. Threads	5	4½	4½	4	4	3½	3½	3¼

WHEAT HARVEST CALENDAR

January—Australia, New Zealand, Chile, Argentina.

February and March—Upper Egypt, India.

April—Lower Egypt, India, Syria, Cyprus, Persia, Asia Minor, Mexico, Cuba.

May—Texas, Algeria, Central Asia, China, Japan, Morocco.

June—California, Oregon, Mississippi, Alabama, Georgia, Oklahoma, North Carolina, South Carolina, Tennessee, Virginia, Kentucky, Kansas, Arkansas, Utah, Colorado, Missouri, Turkey, Greece, Italy, Spain, Portugal, south of France.

July—New England, New York, Pennsylvania, Ohio, Indiana, Michigan, Illinois, Iowa, Wisconsin, Southern Minnesota, Nebraska, Upper Canada, Roumania, Bulgaria, Austria Hungary, South of Russia, Germany, Switzerland, South of England.

August—Central and Northern Minnesota, Dakotas, Manitoba, Lower Canada, British Columbia, Belgium, Holland, Great Britain, Denmark, Poland, Central Russia.

September and October—Scotland, Sweden, Norway, North of Russia.

November—Peru, South Africa.

December—Burma, New South Wales, Northern Argentina.

Page from old-time grain milling catalog.

Courtesy of Robert Beale

How Gristmills Developed

Often a device called a *damsel,* or *millclap,* kept the wheat in the hopper feeding steadily downward, by tapping constantly on the spout. This was said to make a peculiar sound which could be heard above the noise of the machinery of the mill, often called the "damsel's song." Shakespeare in Richard III made a use of this as a comparison when he said "...to have a tongue like a millclap."

The millstones were usually enclosed in a wooden box-type arrangement, called a vat. The flour fell to the inside bottom of the vat and a small scoop-like piece of wood attached to the runner, scraped and pushed the flour along so it fell down below into a sack or bin. In some mills, the flour fell against a board enclosing the sides of the stones, and was scooped up by hand into a sack or bin. The scoop was of wood to lessen danger of fire by sparks. Fires in mills were easily started because of the explosive build-up of flour dust.

Gristmill buildings were ordinarily tall, with three stories and sometimes more, needed for the various operations, and for storage. The wheat was first carried to the top floor for processing and gravity played a part from then on in the mill's operations.

Even later when gristmills were more automated, much the same arrangement was still followed. The wheat went to the cleaning apparatus on the top floor, then down into the grinders in the next floor, filling the hoppers feeding into the top runner stone. It finally reached the bottom floor as flour where it was sacked and taken away by the farmer.

If a miller could afford it, he put a bolting machine in the mill. This was for sifting the flour to remove the bran and shorts. Early bolting apparatus consisted of two or three silk-covered frames which would vibrate back and forth and sift the flour through the silk meshings. This produced a whiter, lighter flour.

Flour ground through the millstones, if not bolted, contained all the parts of the grain, including the wheat germ, and was a cream-colored flour. This was a healthful flour which made delicious bread, but it could become rancid if not used up rather quickly because it contained the wheat germ, the oily part of the grain.

A miller had his problems. There were always the stones needing sharpening. He must watch them when grinding for they needed 'tending' constantly so they could not become too close to each other. He was ever on the alert for fire. As the stones ground out the flour, clouds of white flour dust would float up from the stones, and settle not only on the miller or anyone else in the room, but also collect in all parts of the mill. The mice particularly loved this abundant provision of food for them, becoming fat and prolific. Not satisfied just with wheat and flour, they added a little roughage to their diet by nibbling on the wheat and flour sacks. It must have

Florence E. Sherfey

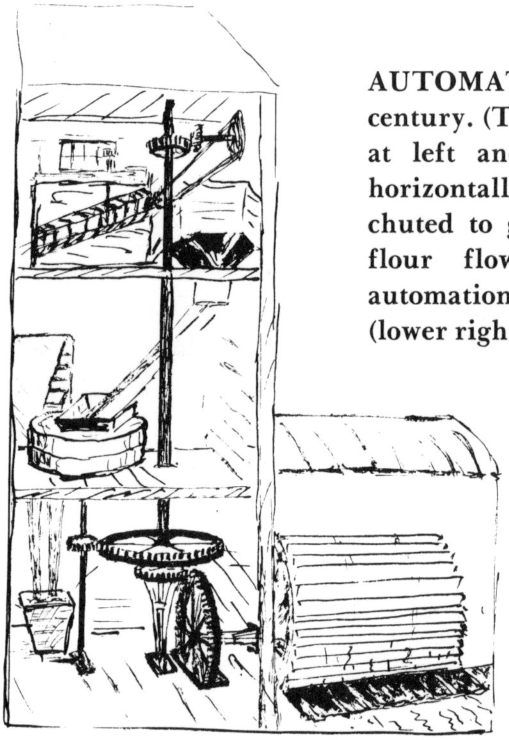

AUTOMATED WATER MILL of the 18th century. (Top), Bucket elevator lifts grain to crib at left and screw driven conveyor moves it horizontally to the hopper at right. Grain is chuted to grinders (center). After it is milled, flour flows to bins (bottom). Power for automation comes from rotating wheel in stream (lower right).

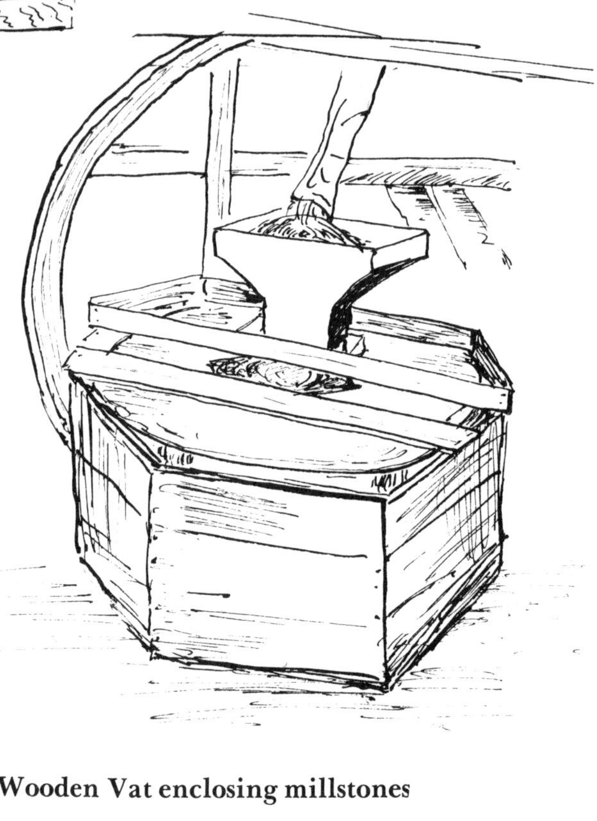

Wooden Vat enclosing millstones

FROM WHEAT TO FLOUR

Flour was made from wheat in a lengthy and time-consuming process. The grain entered the mill by the back door on the first floor, and the Miller made an agreement either to buy the grain outright or to make a sort of trade, such as "I will grind your wheat for one-eighth toll." In other words, he wanted so much wheat in exchange for grinding.

After the Miller had taken his toll, the grain was placed in the wheat hopper over the French Buhr Stones (cleft set).

To better understand what would take place next, examine the diagram to the right. Water entered the mill through a flume box, hitting the water wheel slightly above the middle point. The wheel turned in a counter-clockwise motion, releasing the water at the bottom of the wheel. The water wheel was connected to the master gear (housed in the gear pit on the main floor) by the long, wooden main shaft. The master gear, like the water wheel, turned in a counter-clockwise movement.

Power for the two sets of millstones was derived from the master gear. On either side of the master was a latern gear, which rotated on a shaft perpendicular to the main shaft. The latern gear thus turned the power gear located on the same shaft. At the top of the power gear was a small gear called the hub gear. It was attached to the main drive shaft.

The drive shaft entered the millstones from the bottom stone and was fastened securely in the top stone by the rynd. The bottom stone was stationary and, therefore, the top stone was the only one that turned.

Sitting in the casing of the stones was the grain hopper. Grain entered the stones at a controlled speed maintained by the shoe, passing through the eye into the stones. The stones turned slowly so that the flour was not burned. As they turned, the ground meal worked its way out into the casing and down a shoot into a container on the ground floor.

Thus far the grain had traveled from the hopper to the stones and then to the meal bin.

Flour was the next step. Up from the meal bin via conveyors to the second floor went the shorts, the middlings, and the bran for the final process of flogging. Flogging was a type of sifting done with a cylinder-shaped device covered with a fine woven cloth. Ground meal entered the flogger, which went around and around separating the middling and bran and allowing the sifted flour to fall to the bottom of the flogger into the worm gear. The worm gear worked the sifted flour out to a conveyour that allowed it to fall to the first floor for packing.

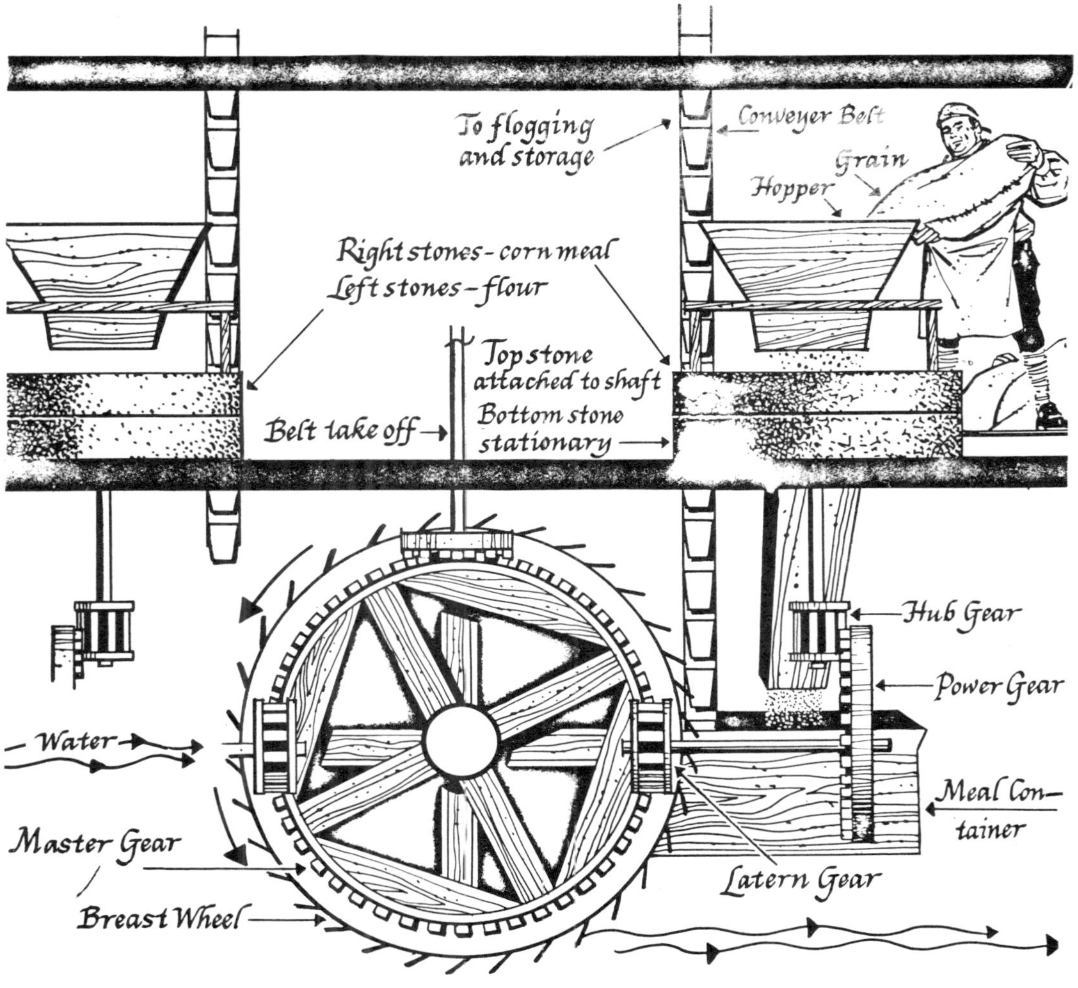

Courtesy of, and used with permission of Commonwealth of Virginia, Division of Parks.

been one of the miller's sweeter joys of life to pick up a sack and discover it contained not a single hole. Most of the sacks had to be patched constantly, or reinforced, to hold the flour.

Around 1860-65, a new type of high protein wheat came down from Canada. To grind this wheat more efficiently a New Process of grinding was developed. This process, adapted from a European technique, was first put into operation in the United States by Edmund La Croix in Minnesota in 1865, and called a middlings purifier. This method was later adapted to roller milling. Rather than working all the wheat berry into a powder in one grinding, a gradual reduction broke up the wheat, regrinding the coarse material several times with the use of air currents and sifting between grindings to remove the bran and middlings. Flour produced in this way was called Patent flour, a name still applied to more refined grades of flour.

The introduction of the roller mill system revolutionized milling. In 1588, the roller mill was first mentioned in a publication by an Italian, Agostino Ramelli. In 1602, G.Z. Bockler, a mechanical genius, developed a mill using two corrugated rollers together with an agitating device for sifting the grindings.

Although rollers had been used to roll metal and process sugar cane, they were not again used in flour production until 1833. Then, Jacob Sulzberger, a Swiss millwright, used them in his flour mill, the method quickly taking hold in Europe, especially around Budapest, giving the name "Hungarian" to the system.

In the United States, an all-roller mill was constructed and operated briefly in Philadelphia in 1876. The first important commercial use of them was in Minneapolis in 1878.

The faces of the steel rollers are patterned with parallel, equally spaced breaks and grooves with sharp edges for cutting the grain. This corrugated pattern is different for each set of rollers, usually two to a cabinet, each pair performing a special function in the gradual reduction of the wheat to flour.

Roller mills appealed to the millers who had been using millstones, because use of them eliminated the bother and cost of dressing the stones to keep them sharp. The process with the rollers was done automatically, a system of belts and pulleys—in the older sets—activating the roller equipment. The product resulting was a more uniform, cleaner, whiter and softer flour which housewives often preferred to the heavier flour from the stones.

But the roller milling also hastened construction of the larger, more efficient mills in bigger cities, which forced the abandonment of the small community mills, especially if they still used only the stones for grinding flour. Often stones and rollers were used in combination in the small town mills, which was a satisfactory arrangement for those wanting a choice of either type of flour.

How Gristmills Developed

Many mills after being built using the water wheel for power, soon felt the necessity of changing over to the rollers. They closed down their gristmills long enough to make the new installations, often enlarging the mill and adding another story, as well as additions. They usually discarded the millstones, glad to be rid of them, and put in four or five of the roller cabinets. The water wheel was no longer needed for a turbine was installed now which still used water, but gave more power and was more efficient. A cleaning system for the wheat was added to the equipment and a complete bolting process installed. Now the mill became a Merchant Mill as well as a Custom Mill, grinding for the business trade as well as for the individual farmers. The flour mill, always the most important industry in the town, now brought even greater benefits for it often gave extra employment and brought added business to the merchants by drawing in more farmers bringing their wheat to the mill.

Those gristmills which did not change over to the new roller system suddenly found themselves neglected and deserted. Farmers, preferring the new white, fluffy flour from the roller mills, began taking their wheat to the larger mills in exchange for the mill's flour, happy also to have the bran and shorts for stock feed.

By 1886, railroads were coming to rural regions, suddenly bringing a new and easier mode of transportation for country folks. Farms were gradually adding new pieces of machinery which made farming easier. The old was giving way to the new.

But even in the 1890s, some gristmills were still being built. Before the 1900s, the Touchet Valley near Walla Walla claimed seven gristmills in a thirty-mile span, each mill running to full capacity in its grinding of flour. Many mills in Whitman County were spaced almost as close, but each also was functioning strongly. Most of these mills, however, operated for only a few years after being built.

Even the mills that had changed over to the roller equipment, as well as new ones equipped with them when built, bought only a few extra years of life. By the early 1900s, the flour milling industry was well along in a quiet little revolution of its own begun when flour was first processed to a whiter product. The roller mills further improved this process, until the refinement resulted in mills of a new, great industry. Hundreds of patents were issued for mechanical purifiers, sifters, cleaners, dust collectors, grain washers, bleaching equipment and other milling machinery, until the small mill found it could not compete. It was "squeezed" out by the larger mills in bigger cities that could hold their own in competition. Many of these mills became large corporations.

And what became of the old gristmill which never changed to the new, but still hugged its ancient machinery to its bosom and stoically remained the way it was born? Most, idle and useless only a few years after they were built, were rudely torn down. One or two old weathered gristmills might be thought to have hid themselves quietly by a moaning brook and silently counted the days left. Perhaps some could even be believed to have committed suicide, going the easy way by fire so prevalent in those days in gristmills with the explosive build-up of flour dust and heat.

Some gristmills survived in the eastern part of the United States. But in the West, where gristmills were dying even as they were being born, only a weathered structure here and there stands, and these long ago gave up their granite millstones for the roller process. Occasionally, an idle old mill building, privately owned, still looms proudly against the skyline, holding its place in the household—and community—like a revered, older family member.

But for most, only a shell, or at best a row of rock and concrete footings marking the location, are all that remains of yesterday's world of vanished gristmills.

Vanished Gristmills

Stevens County Gristmills

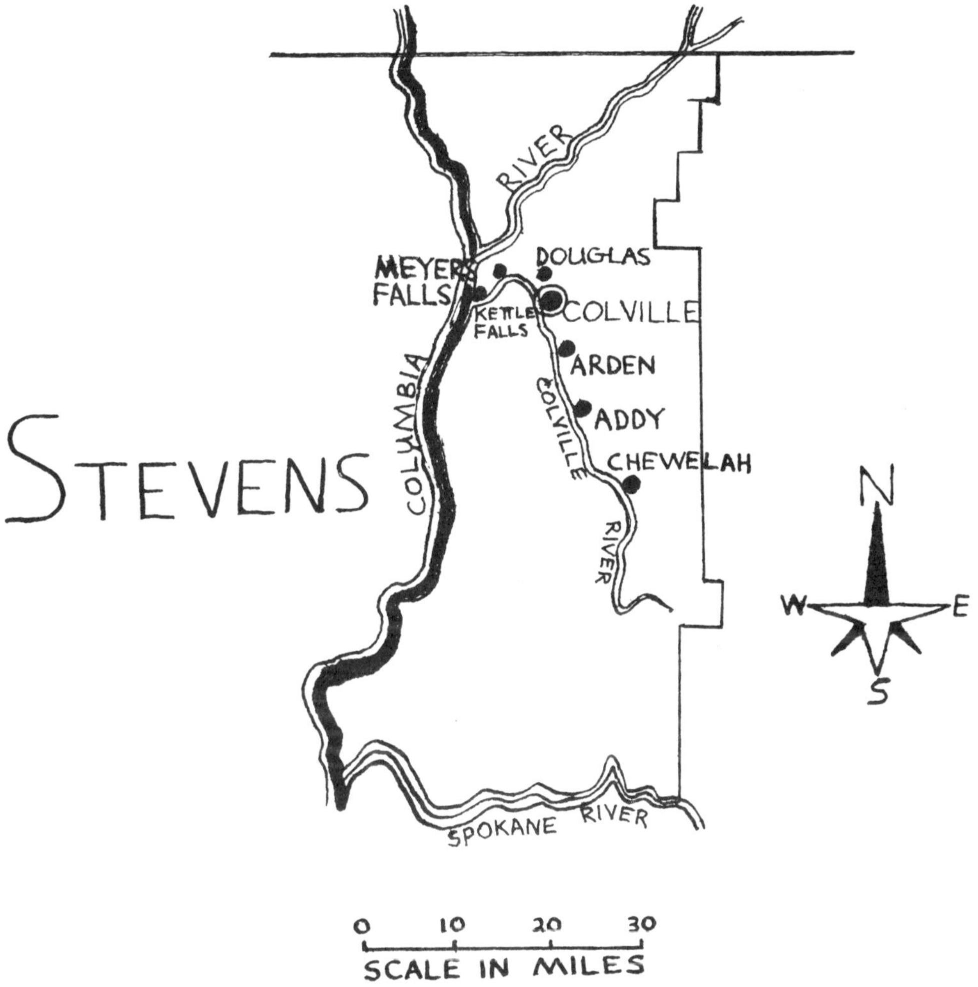

GRISTMILLS IN STEVENS COUNTY

Northwest Fur Company Gristmill
Hudson's Bay Gristmill, Ft. Colville
Meyers Falls Gristmill
Oppenheimer Gristmill, Arden
Douglass Gristmill
Addy Mill
Chewelah Flour Mill
Adams Gristmill

Vanished Gristmills

STEVENS COUNTY GRISTMILLS

Northwest Fur Co. Gristmill
Hudson's Bay Co. Gristmill

MEYERS FALLS MILL

EARLY flour milling in Colville was accomplished by the work of three gristmills. These were built and operating in different periods of time, with all of them in approximately the same location. The first mill was said to have been at Marcus Flats built in 1816; the second, the Hudson's Bay Company Mill, which was called the Goudy Mill, built around 1843, and the third, the Meyers Falls Mill, built in 1872.

The three gristmills were located on an important point of a river where crashing falls drowned out the rumble of granite millstones. The falls were those of the Colville River, now known as Meyers Falls and located about three miles from Colville.

The first two gristmills were British owned. The 1816 mill was said to have been first built by the Northwest Fur Company employees near Marcus, and moved to the point on the falls in 1821 when that company was purchased by the Hudson's Bay Company. Wheat was grown by the Northwest Fur Company employees in 1812 with seed brought in by the Northwest Fur in 1811. (Information and notes on early milling, J.J. Mulvey, and notes on Colville, W.P. Winans.)

Quoting from "*History of Northern Washington*," Steele and Rose, 1904, "History connected with this romantic spot, dates back as far and possibly farther than any point in Stevens County. On this subject earlier pioneers of the county differ. Many reliable sources state that the Northwest Fur Company established a post and built a gristmill in 1816 which they operated for fourteen years. About 1840 it was torn down and a new mill erected in its place. November 19, 1866, L.W. Meyers who had come earlier from Eastern Canada, secured control of this second mill and operated it until 1872."

Stevens County Gristmills

The second mill built in 1843, as most sources state, was called the Goudy Mill, named so after the man who built it for the Hudson's Bay Company and operated it after it was built. This mill replaced the old 1816 mill.

Mr. W.P. Winans, who lived in Colville in later years and was associated with the Goudy Mill under Mr. Meyers as bookkeeper during the 1860s, states in his diary: "This second mill, the Goudy Mill, was built around 1843 by the Hudson's Bay Company since the first mill built around 1816 had probably gone to decay. The Goudy Mill was remodeled in 1866 by John Houser. In 1872, Meyers tore this mill down and built a new one. This last mill is said to have burned down in 1914."

Further information comes from the *Statesman Index*, Colville, October 8, 1897: "At Meyers Falls on the Colville River the Northwest Fur Company erected a gristmill in 1816 only eleven years after Lewis and Clark had completed their memorable expedition. The old burrs (millstones) of this mill are yet on the ground. Agents of the company, however, were in this vicinity as early as 1809, but merely for the purpose of purchasing furs of the Indians making no attempt at settlement. Whether this original mill was torn down or destroyed by fire is not definitely known, but in 1843 another mill was erected on the same spot. In 1863, L.W. Meyers, for whom the falls were named, a Canadian and the pioneer of Stevens County, took a lease upon the building. Pending a settlement of the Hudson's Bay Company with the United States government for the relinquishment of their lands, Mr. Meyers afterward secured possession of the mill and water power and falls—one of the most valuable sources of water power in Eastern Washington."

This comes from W.P. Winan's diary: "When I first saw the (Goudy) Mill in 1861, it was a hand-hewn building about 30 X 50 feet, two stories, with an attic which had a roof covered with cedar bark. It had a single pair of stones made of the granite of the neighborhood. A homemade fanning mill was the only wheat cleaning machinery, and the bolt to separate the flour from the bran was a wire screen, stationary, with a revolving brush on the inside."

The "pair of stones made of the granite of the neighborhood," could be the two stones at the Cheney Cowles Memorial Museum in Spokane. The stones would have been removed from the Goudy mill in 1872 when it was torn down, and a new one built to take its place by Mr. Meyers.

The old Goudy Mill saw many years of use grinding wheat or other grain for farmers and the many Indians in the area. The missionaries from Tshimakain, near Spokane, brought their grain there to have it ground. Reverend Elkanah Walker came often. He and Reverend Cushing Eells went to the Spokane Tshimakain in 1838 as missionaries, and soon after began the trips to Colville and to the gristmill. The mill was evidently a busy place, for Reverend Walker's diary often mentions that

These millstones from the Goudy Mill built by the Hudson's Bay Company in 1843. Stones are hand-hewn of native granite. Left stone is *nether* stone, or bottom, and right stone is *runner* stone, or top one. These are the grinding surfaces showing, with furrows still visible.

Courtesy Eastern Washington Historical Society.

he had to wait, for the mill was busy. He said it took him fourteen hours to go to Colville to the mill from Tshimakain, "walking the horses," and "not including stops by the way." The horses averaged five miles per hour.

These were the days when the wheat crops were laboriously planted and harvested by hand labor. It was broadcast at planting time, and cut with a hand scythe at harvest. Then the grain must be separated from the heads by "flailing." Anything heavy enough to beat but not smash the heads was used in this operation for extracting the grain. Still the wheat was not yet ready for the granite millstones to grind into flour, for there was about as much lose straw and dirt mixed in with the grain as there were kernels of wheat. It was "winnowed," often by employing the wind to help, for a breeze could catch the lighter, unwanted particles and whirl them away, as someone poured wheat downward from a high location.

Before Mr. Meyers took over the gristmill in 1866, and during the years it still was under the Hudson's Bay Company's jurisdiction, many officials of that company visited the mill. Dr. John McLoughlin was the most important and most popular

Hand cut millstones said to have been used in first gristmill built in 1816. (Grooves retouched in photo.)

visitor. He was in total charge of all Hudson's Bay Company posts in this part of the country, and would occasionally come up from Fort Vancouver where he was Chief Factor. The botanist, David Douglas, a studious man who knew every plant's characteristics, and for whom the Douglas fir was named, sometimes dropped in to sit and rest a few minutes.

Archibald MacDonald came into the mill quite often. He was a genial, well-liked individual with whom everyone wanted to talk and visit. He was the Chief Trader at Fort Colville in charge there, and got along well with everyone.

In 1847, after the violent upset at Waiilatpu, with the tragic massacre at the Marcus Whitman mission, the scare unsettled the Spokane mission and because of this, Tshimakain soon disbanded. The missionaries and their families went to Fort Colville and lived there until plans materialized for them elsewhere.

Colville itself, one of the first settlements in Eastern Washington, is packed with historic interest. This began early in the 1800s or even before, with the advent of explorers, fur traders and trappers, as well as itinerant travelers, sometimes mentioned in historical writings by obscure, early writers. Lewis and Clark traveled through this northwest region in 1805 and David Thompson was often there around 1811 or before. The area was constantly under scrutiny by the competing fur companies, each ready to step in and snatch control of the thriving markets created by the coveted fur pelts.

Hudson's Bay Company, the strongest of these competitors, absorbed the lesser Northwest Fur Company, merging with it in 1821. The Hudson's Bay Company also inherited Northwest Fur Company's problems along with its interests, and one of its headaches was the so called *Spokane House*.

Vanished Gristmills

Kettle Falls and Marcus Flats in Colville Valley
(Courtesy Washington Water Power)

Under the Northwest Fur Company, for which Spokane House had been its local headquarters, this establishment had become an expensive operation. This came about mainly from the temptations it offered in high-living for its occupants and especially its "guests" connected with fur trading, constantly flocking to its doors eager and ready for entertainment. Its inland location also made it difficult to maintain, for supplies had to be hauled in at great expense.

One of the first acts of George Simpson, Governor of Hudson's Bay Company, was to order Spokane House disbanded and removed to a new location. The new location was near Colville because the waterways of the Columbia River and the power-attractions of the Colville River, would lessen all transportation expenses. In every way Colville would be a more central location for the new surviving company now under Hudson's Bay Company's command.

By 1825, the move to Fort Colville was achieved. Simpson named it Ft. Colville in honor of Andrew Colvile, a London governor of the Hudson's Bay Company. "This was a wooden fort of large size enclosed with stockade and bastions. Buildings were made of cedar logs and cultivated fields surrounded the fort. The fort was considered next in importance to Fort Vancouver."

There was more than one fort near Colville at different periods; the Hudson's Bay Fort established in 1825 and a military fort established in 1859 by Major Pinkney Longenbee, which was abandoned in 1881. (H.P. Isaacs of Walla Walla is said to have built a gristmill at the military Fort Colville in 1859.)

Some old timers state that Marcus, or a part of it was a fort around 1816. It might be said that most any little settlement in all parts of the state, was a fort at one time or another in those early days, fortifying itself against Indians prevalent then. Hundreds of Indians had been gathering at the fishing grounds at Kettle Falls since prehistoric times, and a tiny white village could have been at their mercy if the Indians had chosen to band against them. At Marcus, one can assume a gristmill existed, as mentioned, although somewhat primitive. A sawmill was operating there, too, with the possibility that both gristmill and sawmill ran as a combination. The remains of the open pit for whip-sawing remained visible for many years.

About one mile from the Hudson's Bay Fort Colville, the Kettle River emptied into the Columbia, churning into it in a pile-up of foaming waters which formed high, crashing cataracts. This was called Kettle Falls, taking the name from the large, round, pot-holes in the rock sides above the river. The falls provided not only the exhilarating sport of capturing salmon for the many Indian tribes who had been coming to this spot since early times, but also gave them their winter supply of food. During the salmon season, Indians from hundreds of miles around came here to camp.

When L.W. Meyers came into the region in 1856, he was anxious to secure property in this wide, fertile valley, especially property along the Colville River including the gristmill. He considered the gristmill a good business venture. He also felt a store built here would be profitable, and later he built one which he operated until 1900.

When Washington became a territory in 1853, the United States took over the property the Britishers had settled on at Colville in 1825. The Hudson's Bay Company now had to withdraw from its location, but the Company claimed thirteen square miles which the United States agreed to buy from them. This included the gristmill on the Colville River falls.

Vanished Gristmills

The old Goudy Mill built by the Hudson's Bay Company in 1843. Notice the construction of the building, long siding boards mortised at the sides. Overshot water wheel may have been underneath the mill, taking water from the river backed up in a large pond, or reservoir.

Courtesy Eastern Washington Historical Society.

Stevens County Gristmills

Section of Kettle Falls River. Falls are a succession of rapids tumbling over the big rocks.
Courtesy Washington Water Power Company

Meyers leased the property in 1860 and homesteaded it in 1863, but it could not be officially his until the government had finished paying for it. The last payment was made to the Hudson's Bay Company in 1872.

Meyers had a partner helping him whose name was George B. Wannicott. It was soon evident that another miller was needed to handle the mill, and they found an excellent man for this work by the name of John Houser. John was an experienced miller who had received his training in that line in Germany as a young man.

He had come to this country from Germany in 1848, seeking employment in eastern flour mills without success. Destiny led him to California and then to Alaska in gold mining ventures, some of which proved fruitul. In 1866, "A number of men came down from the French Creek mines and John Houser was one of them." Learning of the gristmill and the opening for a miller, he applied for employment and was hired in the fall of 1866 as miller.

Vanished Gristmills

West side of Meyers Falls mill, looking East. Gristmill seems to be built on the water but as other pictures show, is well situated on firm ground. Water wheel was probably beneath the mill, in the opening of building seen at end. The falls are the Meyers Falls on the Colville River. Picture 1887. Courtesy Eastern Washington Historical Society

Stevens County Gristmills

East side of Meyers Falls gristmill looking West. Reservoir is in foreground. Water after coming to the mill emptied back into the river on west side of building. Gristmill used one run of handmade granite millstone, with simple cleaning and bolting machinery at first. Later more complicated bolting system installed. Picture 1887.

Courtesy Eastern Washington Historical Society.

John Houser, now thirty-seven years of age, single, intelligent and energetic, was glad to be back in the milling field. The first thing he did with the help of Meyers (an experienced cabinetmaker) was to remodel the gristmill. This was still the old Goudy Mill as described by Winans in 1861, with the two hand-made granite millstones. These stones were retained, but the rest of the mill was worked over and put in better condition. Houser constructed a "middlings purifier," a new process method later adapted to roller milling. Edmund La Croix first used such a machine in Minnesota in 1865. This produced flour made from endosperm, the inner or embryo particles of the wheat, which was freed of bran by a gradual reduction process. Rather than working all the wheat kernel into a powder in one grinding, a miller broke up the endosperm into bran free granular "middlings", or farina. The best grades of flour were made by regrinding this coarse material which passed over a vibrating screen. An upward current of air lifted off most of the light branny particles, or "purified" the middlings. Flour produced in this way was known as "patent," a term applied to more refined grades of flour.

Houser's machine was somewhat primitive, but produced an excellent patent flour, his home made wooden fan for the air current, at first perhaps turned by hand.

Houser achieved another "first" at the Meyers Falls mill by producing the Farina which had not been manufactured before. A wide market developed for this and wagon loads of it were sacked and hauled to Walla Walla where it appeared for sale in the markets. It was also sold to many miners laying in supplies on their way to the Idaho mines. When John Houser's farina and his fine flour were exhibited at the State Fair in Salem in 1868, they won first prize.

Houser also made a device for cleaning the wheat at the Meyers Falls mill, which helped in producing a better grade of flour.

The Meyers Falls mill ground grain for the farmers and the Indians throughout the region continuously for almost ten years. Its flour was of high quality and in wide demand. John Houser constantly upgraded his product, often combining different grades of wheat to create better flours for bread making. Later, he was to specialize and excel in this in the flour he produced at his own mill at Pataha City in southeastern Washington.

The old building was showing wear. The huge old timbers of the Meyers Falls Mill still held as firmly and strongly as ever, mortised and tenoned in the classic manner of that long ago age when nails were not used. But some of the wood itself was decaying near foundation levels where moisture was constant. This decay was encouraged by the thick layers of flour dust built up over the years, absorbing enough moisture to bring chemical changes not conducive to preservation of wood fiber.

In 1872, Meyers closed down the old Goudy Mill, taking a year or two to demolish it and put up a new one. At this time, John Houser returned to Germany for a visit.

When Meyers built the new gristmill, he used some of the old wooden materials from the other building, which were still in good shape. The floor boards, thick, hand-hewn planks, were still sound, as were the huge timbers. He discarded the hand-cut millstones put in so long ago, and it is believed they are the handmade millstones at the Cheney Cowles Memorial Museum in Spokane. They are rough cut and heavy, and must have been tremendous chore to chip out from the native granite found along a hillside. They give appearance of having been able to do their work of grinding grain, however, and that was their purpose.

After the new mill was completed, Mr. Meyers operated it, built in the same location as the old Goudy Mill, until 1889. At that time he discontinued its use.

The mill burned down in 1914. Before that time, many people came to view the falls and visit the gristmill, and learn the story of the old Goudy Mill which had stood in its place for so many years before. Old timers in the Colville region tell stories of the Hudson's Bay Company's Fort Colville and the people who lived there. Archibald MacDonald, the Chief Trader at the Fort was a good friend of people living in the area. The town itself has its own story, being the final settlement that began with Pinkney City, which was three miles to the east of the present city. History gives the country a rich background, and a relating of scraps of it could always have been welcome to visitors of the Meyers Falls gristmill—old itself in 1914.

It is said a careless visitor dropped a burning cigarette ash to the floor on which were some wood shavings. It quickly ignited the material on the floor and before anyone could put it out, the fire was out of control. The flames destroyed the last old mill in Stevens County, located at its picturesque spot at Meyers Falls, on a point of the Colville River.

Vanished Gristmills

OPPENHEIMER MILL

The Oppenheimer Mill originated through the efforts of Judge B.Y. Yantis helping Chief Garry of the Spokane Indians build his second gristmill for the Indians in his region.

Chief Garry had previously erected a gristmill in the early 1850's. This was on land which he then owned, and the site of the mill was near the Spokane River, called Silheim Springs. Later this property was part of the J.P. Glover estate. The gristmill Garry built was widely used by the Indians who came in large numbers with their grist, mainly corn, for grinding into meal or flour.

It seems some early whites in the Spokane region observed the success of this primitive mill, and they shrewdly planned to get in on the operation. They contrived to gain control of the mill under the guise of "helping the Indians," managing the operation so they would reap the benefits for themselves. Before long, they even gained title to the land, finally securing and owning the property.

Rather than allow the whites to encroach in this fashion, the Indians reacted in the only effective way they knew—destroying the mill. One night, one or two of them crept stealthily to the mill and started a fire beneath it. In a short time, the entire wooden structure was a fierce blaze, the mill burning swiftly to the ground. All that was left was a little of the machinery.

B.Y. Yantis, known as Judge Yantis, was prominent in the area as well as in the state's early history. He understood Garry's position and problems and was sympathetic with him. He offered to help him rebuild the mill at a different location on the Little Spokane River where the Indians could more safely use it.

The Judge often took trips to Olympia and he agreed to bring machinery which Garry needed from there. The stones would be ordered from France, arriving somewhat later.

It seems that Garry, in the interim, got "cold feet." He was discouraged and probably distrustful of white men in general, afraid that the new mill would fare no better than the old. Garry was a fine man, a well educated person, taught in the white man's schools and an Indian loyal to his race. But though intelligent, he was not a strong, dominant individual. Often sensing the futility of peace and understanding between the white man and the Indian, he did not always stand up for his rights.

Thus, Garry gave up the idea of the gristmill although by now Judge Yantis was already on his way to Olympia and could not know of Garry's change of mind. When the Judge returned, he had the machinery with him. Although disappointed that Garry did not want to go on with the project, the Judge decided he would build a gristmill somewhere for himself.

The Judge was familiar with the water resources in the Colville region, and especially favored a location on the Little Pend Oreille River, about six miles from Colville at Arden. There he took his machinery in 1858 with plans to put up the gristmill. With the help of a Michael LaFleur, Judge Yantis put in a dam and a ditch for the water. He then secured a good carpenter, Thomas Brown, as the millwright to build the mill and the wooden box-flume which would bring the water to the wheel. The building was built mainly of logs from the nearby wooded hills. It was a small building, only 12 by 14 feet, and constructed without any nails whatsoever, the joints all mortised, tenoned and pinned in the classic manner of that age. The mill was finished in 1859.

The gristmill, although crudely but sturdily fashioned, was also a unique one. Oftentimes, mills in those early days were built entirely of wood, which included all the machinery, and such was the case with the Oppenheimer mill. All the equipment in the gristmill, with the exception of the millstones, was wooden. The wheels and shafts and gears had all been formed and carefully fitted together from the native lumber, and this wooden machinery worked as well as if it had been cast in iron or forged out of steel. The wooden cogs were set separately into the wide edge of the good-sized power wheel. If one of them became damaged or broken, it was easily pried out, like a bad tooth in an aching jaw, and a new one, whittled to fit, inserted in its place.

The Pine City mill in Whitman County had machinery which was made entirely of wood, and sheep's tallow was used on the cogs to lessen the wear. No doubt, some similar lubrication was used on the Judge Yantis Mill.

The granite stones, which came from France, were small, less than ten inches in diameter and set vertically instead of in the usual horizontal position for grinding wheat. The lack of space in the small room was a determining reason for placing them this way, but stones sometimes were put in this position in a few gristmills.

A water wheel of an overshot design powered the mill. The water was brought by way of a wooden flume from the dam and poured down over the wheel from overhead. A wooden shaft from the big water wheel entered the mill connecting to the power wheel with its wooden cogs. These meshed with other wooden gears and turned other shafts ending with gears, one of which locked into, and turned the two millstones. It is said the mill also bolted its flour, after a fashion, but this was a very simple, hand-operated device.

A man by the name of C.H. Montgomery, as Mr. Winans reports in Durham's *History of Spokane and the Inland Empire*, went to the mill one day to get some bran. He found the Judge bolting the product as it came from the stones. He was shaking a small box over a larger one (by hand). The bottom of the small box was covered with cotton cloth like mosquito bar, and the larger box was receiving what went through. "He asked me how much bran I wanted." Montgomery said, "I told him half a bushel, and he replied, 'I can't get that much today.' I watched him a short time and began to think so myself, for the cloth was so coarse that all the siftings went into the larger box with very little bran being collected."

This early mill can be considered to be the first American mill in the eastern part of Washington, since the other early Colville mills were British-owned by the Montreal-based Northwest Fur and the London-based Hudson's Bay Company.

The Judge Yantis gristmill changed hands in 1861. At that time he sold it to a Canadian by the name of Hoag, who in turn sold the mill the following year to D.H. Ferguson & Company. The two partners of Ferguson were brothers by the names of Samuel and Joseph Oppenheimer. Ferguson and his partners made improvements in the mill, first enlarging the building and putting an addition at one end. They also put in a small bolting apparatus and a wooden turbine wheel to step up the output of the mill. They employed a miller by the name of Thomas Martin to run the mill.

Martin with help, removed the small, original ten-inch stones and put in new and larger millstones. These new granite stones came from the hillside nearby, and were laboriously fashioned by hand into good-sized millstones. They were said to have come "from back of Thomas Haller's, right across the valley from Orin, in the hill, and were drawn over to the Pend Oreille mill by wagon and team." Here, when work slacked off a little, the miller with the help of another man hacked away at the big rocks until finally a round, workable shape resulted. Their tools would have been little better than a hammer and a sharp-edged chisel fashioned in a nearby blacksmith shop, or devised themselves. Later, to keep the furrows on the surfaces sharp, C.R. McMillen of Orin was employed and he dressed the stones at regular intervals.

In 1868, Ferguson sold out his share to the two Oppenheimer brothers in his company, and the mill took the name of *Oppenheimer Mill* at that time. According to Winans in Durham's *History of Spokane and Inland Empire*, the Oppenheimers hired John Houser of the *Meyers Falls Mill* at Colville to remodel the mill as he had the Meyers Falls Mill, and install a middlings purifier in it, in the summer of 1868. This was a bolting machine which separated the bran and shorts and produced a white, fine flour.

Stevens County Gristmills

Ancient millstone of the Oppenheimer Mill at Arden. At the time of this picture, 1939, they were still submerged in the mud near the mill site, but have since been removed and placed in a monument-type setting at Colville.

The one shown here, belongs to the two which were chipped out of native rock by Thomas Martin, miller, for the Ferguson Company around 1882. Seated on millstone is P.L. Rockwell, Colville valley pioneer.

The original Oppenheimer Mill as it looked about 1890. Mill was small, 12 X 14 feet, and built by Judge B.F. Yantis in 1859, with the help of Thomas Brown, carpenter.

Courtesy Eastern Washington Historical Society.

Vanished Gristmills

A unique observation comes from a reporter of the Portland Oregonian who visited the mill at one time. In the May 1881, issue of the *Portland Oregonian*, he described the Oppenheimer Mill, which still retained its wooden machinery: "A quaint old structure, made principally of logs and standing in a little corner against the mountains on the bank of the Calispel River. Going inside, the first thing that attracts one's attention is the total absence of that creaking noise and jarring peculiar to other mills. Not a single piece of iron shafting, no pulleys, connecting rods or gearing of any kind can be seen. Everything is made of wood. The shafts and pulleys are large, cumbrous-looking things, but run easily and without noise, and, with an old turbine water wheel, were made by hand from timber cut on the hillside. Even one of the millstones was taken from a bed of granite on the neighboring hill. It has a capacity of about 30 barrels per day, although it is seldom called upon to turn out that amount."

The mill is said to have "flourished for many years" grinding flour for the many farmers in the region and the Indians who lived there. Very little information regarding the mill after the 1880s is available. Stevens County was the center of farming, trading, mining and milling from 1812 until 1880, but by the 1880s other areas and cities were opening up and developing, and flour milling was fast becoming a well-established and leading industry in all towns by then.

George Morley owned the property on which was located the gristmill in the 1900s, and he started the Arden orchards on the farm. Judge W. Lon Johnson purchased the farm of 320 acres in 1917 from a mortgage. In 1918, Judge Johnson tore down the mill, using much of the lumber from it to construct barns on the farm. He later called the farm, *Elm Tree Farm*, because of the 250 Chinese Elms he planted there. The mill dam remained and was used for irrigation purposes.

For many years the site of the mill and dam was a recreation area and a gathering site for all important events, used as a picnic center for the Colville area people. But when the farm changed hands again in 1946, this was closed down no longer open to the public.

The millstones from the old Oppenheimer Mill have been preserved and are displayed on the southwest corner of the Stevens County courthouse block in Colville.

These gristmill mementoes belong to Colville Valley's past, an era of vital interest historically. Gristmills *grew*, matching an area's development, and a gristmill's history reflects the story of that area and its people. Nowhere is this more apparent than in the Colville Valley. Although other gristmills were there first, the Oppenheimer Mill, an outgrowth from an Indian's desire to help his people, can claim distinction to being the first American Mill in Eastern Washington.

Stevens County Gristmills

The Oppenheimer Mill on the Little Pend Oreille River at Arden in the Colville Valley. Established in 1859 by Judge B.Y. Yantis, as an outgrowth of the Judge's trying to help Spokane Chief Garry build a gristmill for his people.

Courtesy Eastern Washington Historical Society.

Vanished Gristmills

STEVENS COUNTY GRISTMILLS

ADDY

A mill was built at Addy in the late 1880s by a John Chapman. Mr. Chapman is believed to have worked at the Oppenheimer mill for a time before starting a mill of his own at Addy. Very little information is available about the gristmill, other than it was not too successful an operation.

DOUGLASS MILL on Mill Creek

In 1859 or 1860, R.H. Douglass built a gristmill adjoining a sawmill which he and John Nelson built, financed by Frances Wolf. The power for the gristmill was provided by "buckets on a fifty-foot endless chain over a three-foot pulley." It is said he called the gristmill, "Love's Defeat."

The sawmill suffered from lack of business because Douglass and Nelson charged exhorbitant prices for their lumber. Although the sawmill had been built by them with the thought of providing lumber for the garrison, the major in charge would not meet their terms and got his material from other sawmills. It is also stated that Douglass claimed the water rights and "froze the others out," which also made him unpopular.

The gristmill was built a few months after Judge Yantis began his mill on the Little Pend Oreille, and since there was not enough business for both the gristmills, the Douglass mill closed down.

The location of the Douglass gristmill and sawmill in later years became a historic site commemorated by a beautiful park-like picnic grounds.

CHEWELAH

A gristmill was built at Chewelah in the late 1860s, its location "north of the present New Congregational Church." The gristmill was operated by water power from the Colville River. A dam was built near the center of what is now the city park and the water from it came to the mill by way of a wooden box-flume.

A company was formed in the Chewelah community to build a $50,000 mill, it is said.

The gristmill was a two-story log building and had two run of millstones, or two sets, one large set and one smaller pair. The larger run of millstones was used for grinding the wheat into flour and the smaller one for grinding corn and cereals.

The mill was under the operation of John McFadden. In later years it was taken over by L.B. Fenwich who owned and ran the gristmill.

Stevens County Gristmills

ADAMS MILL

The *Business Directory and Guide to Washington Territory, 1872* in its *Table of Gristmills*, lists three gristmills for Colville Valley. One was the Pend Oreille owned by Oppenheimer and Company; the Hudson's Bay Mill run by L.W. Meyers and Company, and the *Adams Mill*, run by Arnold Adams.

No other information concerning this Adams mill seems to be available at this writing.

It is known a gristmill was built at Fort Colville around 1858 by H.P. Isaacs of Walla Walla. The Adams Mill could have been this gristmill.

Sketch of the millstones in their monument-type setting in Colville.
(Drawing by Randall Johnson)
Courtesy Glen Adams

Vanished Gristmills

Walla Walla County Gristmills

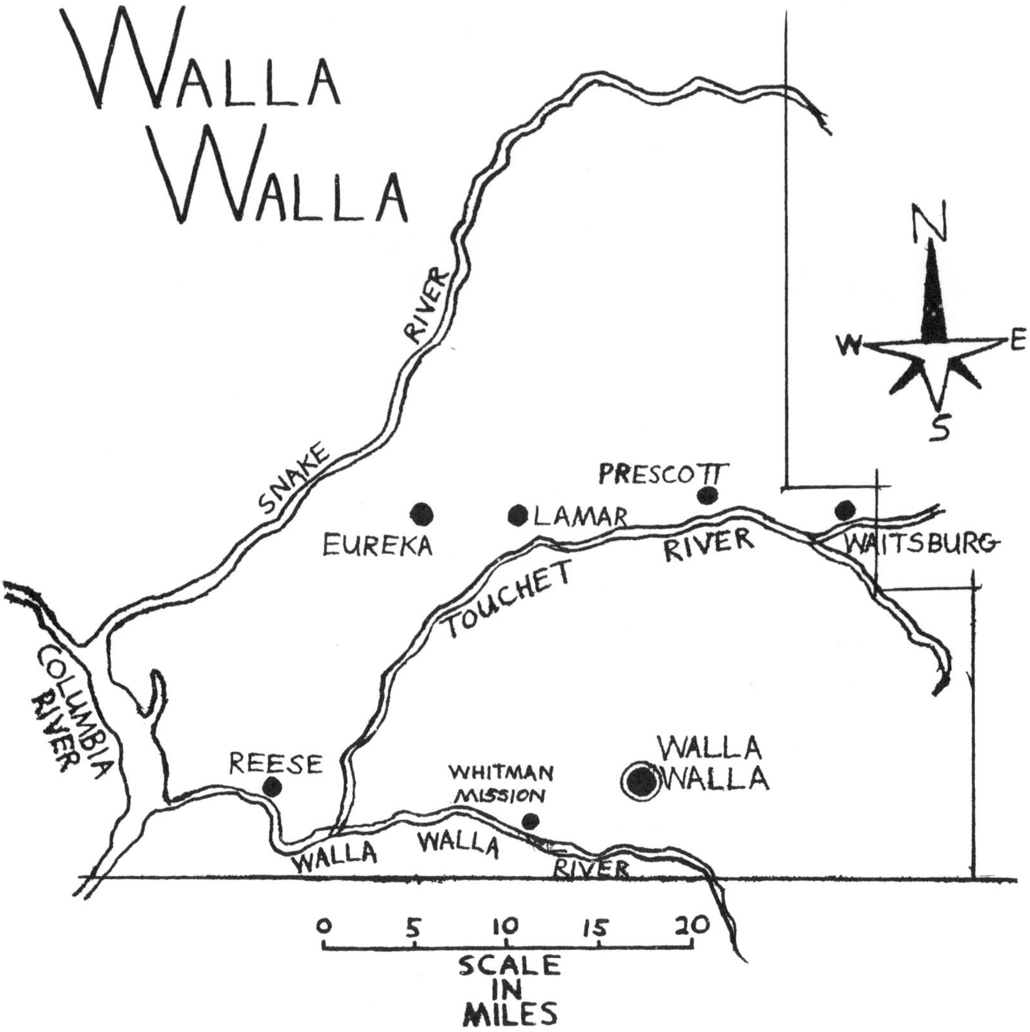

Pioneer Gristmill
Frontier Gristmill
Coyle's Mill
Eureka Mill
Wait's Mill, Waitsburg
Marcus Whitman Gristmill
North Pacific Flour Mills
Prescott Flour Mill
City Mills
McCalley, or Reese's Standard
Acme Mill at Lamar

Walla Walla County Gristmills

PIONEER GRISTMILL

THE first gristmill in Walla Walla County was the *Pioneer Mill.* It was constructed in the year 1859 on the Yellowhawk Creek[1] by Almos H. Reynolds.

Almos, or A.H. Reynolds as he became known, was an exceptionally fine millwright as well as miller. He came from the East, New York, well-trained in this line of work. His father, Nicholas Reynolds, was also an excellent millwright and miller and no doubt he must often have hoped that Almos and he might eventually go into the milling business together, the two of them owning and running their own mill. But this was not to happen, for Almos could think of little else beyond his school work and his apprenticeship in a nearby mill, than to join his friends and neighbors who were going westward.

Almos was still a very young man when he said goodby to his parents for the West. They were gratified knowing their son would never go hungry or be without work with his industrial background in the milling trade.

Their son never did go hungry. No matter where Almos stopped he was offered an inducement to build a gristmill, if he cared to remain in one region long enough to erect one. He first got as far as Illinois, then to Iowa, stopping for a while in each, working at his trade for extra money. Later he moved on to the Pacific Coast which was his destination, reaching the Willamette Valley in Oregon where he settled down for a short time to live.

But there was sadness already in the young Reynold's life, for he had married a young woman before coming to Oregon perhaps on the way West. It is known that his wife and child were killed before he crossed the mountains. But after this tragedy he went on, putting his troubles behind him and working as hard as he could in Oregon constructing water-powered gristmills. He no sooner finished one mill than

1. The Yellowhawk Creek may also have had an earlier name: *Panka Creek.*

he was sought out to build another. He constructed mills in Benton and Polk counties and he built the first woolen mill at Salem.

In the summer of 1858, he left Oregon to come to Walla Walla. He took passage on the steamer *Colonel Wright* making the scenic trip up the Columbia River and landing at Wallula, at which place was then located the old Fort Walla Walla. From there he moved on to Walla Walla by stagecoach. He must have been disappointed at the size of the small, quiet settlement of Walla Walla near the Blue Mountains, after the bustling towns of the East and even some in Oregon. But Walla Walla itself was just on the threshold of a sudden expansion, for in two years when gold would be discovered in Idaho, hordes of miners would pass through this town, stopping to stock up with supplies to take with them.

At this time, 1858, Walla Walla's main business was supplying flour and whatever other merchandise or necessities which were required, to the army people located at the fort in Walla Walla. This fort was also called Fort Walla Walla, erected there in 1856 after the Indian war.

Flour especially, was in great demand at the Walla Walla fort but it was difficult to obtain since it had to be brought in the long distance away from Milwaukie, Oregon, in the Willamette Valley by oxen, or horse teams. There was no gristmill closer which could provide them with flour. Naturally, the cost of hauling this flour not to mention the time involved, was so excessive and took so long it had become imperative something must be done to remedy the situation. A good deal of discussion had taken place among interested men in the town and at the fort about constructing a gristmill at Walla Walla or close by, but there was hesitation about beginning one because there were only two or three farms in the region raising wheat, although crops from these farms were excellent.

A.H. Reynolds' arrival in Walla Walla was a timely one and he was immediately approached and consulted as one who could offer advice as well as construct and run a gristmill. After several meetings and discussions, Reynolds decided to build a mill, and two other gentlemen, John A. Sims and Captain Dent then of the army, offered to go into partnership with him furnishing their part of the capital. The selected location for the mill was on a level bank of the Yellowhawk Creek, "two miles south of the town of Walla Walla and one and a half miles south of the fort."

Reynolds' skill as a millwright was apparent as soon as his building began to take shape. The huge timbers used for supports and bracings throughout, hand-hewn in the manner of the day, were meticulously fitted. Beams were firmly joined for strong holding in snug mortised and tenoned bracings, these joints further secured by being pinned together with strong, tapered wooden pins, usually of hard

Walla Walla County Gristmills

A.H. Reynolds, early millwright and miller in Walla Walla County. Mr. Reynolds was also an active civic leader doing much to further development of Walla Walla. He built the first gristmill in Walla Walla.

Courtesy Whitman College, Penrose Mem. Library

wood, tapped firmly into them. No iron nails were used, or even available then, nor needed, for buildings erected in the manner of the early gristmills could sturdily withstand the destruction of time and remain still in good condition a hundred or more years after being built, although the builder may not have constructed his building then with that intention in mind.

Men skilled in millwrighting, such as Reynolds and many other men who constructed mills in Eastern Washington from the 1850s to the late 1880s and '90s, were artisans of their craft. Their names are spoken of in terms of admiration and respect by all who knew these men, or of them and saw their work: Reynolds, Wait, Stimson, Isaacs, Snyder, Bauman, Metzger, Cunning, Dart, Post and others, all peers in their trade.

It must have been as important to a millwright of the 1870s that he measured up as a worthy craftsman in another millwright's estimation or a fellow workman's eyes, as it was to know his work might survive time and endure for one hundred years or more.

The mill on the Yellowhawk was finished in 1859 and called the *Pioneer Mill.* It was well able to withstand all the shocks and vibrations from the heavy, wooden water wheel which Reynolds also constructed, and bear the enormous weights of the huge granite millstones. There was one run of granite stone, each stone 36 inches in diameter. They were imported from France, shipped 'around the horn' like all freight sent to this part of the country then.

Millstones from France, and later from the eastern part of the United States when it was discovered granite there was of similar high quality as that imported, often came in sections. This consisted of eight or nine pieces, cut to fit together in the enclosure of a wide, eight to eleven-inch iron band encircling the stones and holding them in place.

Much, if not all of the remaining equipment in the Pioneer Mill could have been made of wood, including the large power or master wheel, the gears and the big water wheel and its shaft, all of wood. Water to run the mill was collected in a small dam and carried to the overshot water wheel by way of an elevated wooden flume. There was probably no bolting or cleaning equipment, for this first Walla Walla mill undoubtedly was of very basic design, with only the grinding equipment of stones.*

*It is believed that Reynolds secured an old pair of small granite millstones from the Marcus Whitman site, which he installed and used for grinding chop feed. (Nat'l Archives: Nat'l Park Serv. *"Marcus Whitman Historic Site Report," 1973.*)

Walla Walla County Gristmills

THE OLD PIONEER MILL

This was the first gristmill built in Walla Walla and constructed on the Yellowhawk Creek about a mile and a half south of the old Fort Walla Walla (Vet. Hospital grounds). It was built in 1859 by A.H. Reynolds in partnership with John Sims and Captain F. Dent, a brother-in-law of Gen. Ulysses S. Grant. A.H. Reynolds dissolved partnership with the other two men when they decided to use a portion of the building as a whiskey distillery to supply whiskey to the mining camps. Later, a pork packing department was also added to the mill enabling the owners to provide flour, bacon and whiskey to the miners; these three items said to be the basic provisions in food for a prospector. This building, used as a barn for many years, was torn down in 1940.

Courtesy Whitman College, Penrose Mem. Library

Vanished Gristmills

There was instant demand for the flour which the Pioneer Mill produced. Suddenly the Walla Walla country underwent a marked change finding itself used as a steady pathway by the number of miners traveling to the Idaho mines. This little town was unprepared for such a deluge of business and until it became adjusted it responded in the only fashion possible—high prices and often complete lack of the product from over-demand. The flour from the mill was very popular, and also in good supply, but if a miner couldn't pay for a hundred-pound sack of the precious white powder he usually didn't get it. Sims' name at the mill has gone down in histoıy as being one man, the only man, to give credit to the Pierce outfit, letting Captain E.D. Pierce in 1860, have one thousand pounds of flour on credit. In this instance, Sims' judgment was sound for he was paid the following spring in gold dust when this outfit "struck it rich."

The Pioneer Mill soon found another lucrative market for its product; the stores in the eastern part of the United States, for in New York, their flour sold at a price of twenty-five cents a barrel higher than any other flour in that market.[2]

In 1861, two years after the mill was finished and operating, Reynolds "dissolved partnership" with Sims and Dent. Then, on his own, he started building another mill "in competition" to his former partners. There was undoubtedly some disagreement over matters relating to the running of the mill and this could have been caused by Sims and Dent wanting to enlarge their business, not to make more flour, but to convert a portion of the mill into use as a distillery, which they did. A.H. Reynolds with his strict New England upbringing in religious principles, undoubtedly could not sanction the idea of being associated with the manufacture of alcoholic beverages.

It is said that the main items in the miner's supply bags were flour, bacon and whiskey. Sims and Dent were anxious to cash in on this brisk source of revenue. Dent's sister was married to Ulysses S. Grant, at that time a general, and Grant lent his brother-in-law money to enlarge the mill for the distillery. The 'rich bran and shorts were particularly well adapted for distillery purposes.'

It must also have seemed a shame not to go the whole way, having gone this far, so they added yet another department to the industry, a pork-packing business, and now Sims and Dent were well able to take full care of all the miner's wants; flour, bacon and whiskey.

A man by the name of Overholtzer handled the pork-packing part of the business for Sims and Dent, and before long, Overholtzer in conjunction with two men by the names of Jacobs and Scott took over the business, buying it from them and continuing its operations, including the flour making.

2. (New York Tribune), 1868: *Union Bulletin*, Walla Walla.

Although some state a separate building was erected as a distillery, most accounts claim that this business was carried on in a section of the old flour mill, the Pioneer Mill, with an enlargement added to it. In an account written in 1908, by E.S. Isaacs, son of H.P. Isaacs, early mill builder and operator, he states: "...and in response to the demand from Idaho mining camps a few years later there was added to the enterprise an extensive pork-packing department and a distillery...the old Pioneer Mill building with its massive frame of hewn timbers is still standing (in 1908)."[3]

Further historical statements mention that the Pioneer Mill was built on land owned by Jesse Drumheller and later by Dr. Thomas Campbell and now part of the Whitney place, and that after it was discontinued as a mill, the building was used as a distillery, and later under Mr. Whitney's ownership, as a storehouse.[4]

The old mill on the Yellowhawk passed through a colorful era of Walla Walla's history. It was run by a serious-minded young man who held to the principles of honesty taught to him in his youth. He never returned to New York but was in touch with his relatives and friends there often. The death of his first wife and child so early in his life was a tragic experience and he did not marry again until many years later. Reynolds' part in the development of Walla Walla was furthered by his interest and involvement in many civic matters and he took an ever increasing part in the community life and its welfare as time went on, aiding in Walla Walla's growth and development.

3. Walla Walla *Union Bulletin*, June 6, 1948.
4. *History of Walla Walla County*, 1901, Gilbert

Vanished Gristmills

WALLA WALLA COUNTY GRISTMILLS

FRONTIER GRISTMILL — Also called, *STAR*

The second gristmill built in Walla Walla County was the *Frontier Mill*. It was on the Yellowhawk Creek and constructed in 1862 by A.H. Reynolds, who built the first mill also, the *Pioneer*, in 1859. The Frontier Mill, which later took the name of the *Star Mill*, was designated in later years as situated "about where the highway crosses the stream south of Prospect Point School."[1]

When Reynolds dissolved partnership with Sims and Dent at the Pioneer Mill in 1861, Reynolds immediately turned to the building of another mill, the Frontier. Although no motive has been given in historical accounts for the break in partnership, it would seem that Reynolds' pulling away might have been caused by his partners' wanting to incorporate a whiskey distillery business in the flour mill, and produce whiskey in addition to flour. Reynolds may have been adverse to having his name linked with liquor production, and since he was in the minority, he withdrew as the third partner.

Reynolds was an experienced millwright and miller. He had been trained in this trade as a young man in the East where his father also was a millwright and miller. By now, Reynolds, well established in his work, must have had a substantial bank account, large enough to handle the building and running of another gristmill, which he began right away, and he would have been well able to construct it in the way he wished, without partners..

The new Frontier gristmill when finished in 1862, had one run of granite stone, imported from France, and a capacity of producing 50 barrels of flour daily. There was plenty of water in the Yellowhawk and a dam supplied the water for the overshot water wheel.

Reynolds employed other millers to help him in the mill, for there was plenty of work there for more than one man. One of these was Andrew McCalley, a very capable miller of those early times in Walla Walla. He was with the Frontier Mill until 1872 when, like many millers then, anxious to take on and operate a milling business of their own, he bought into a mill west of town and left the Frontier, or Star Mill, which it may have been called then, to run his own mill.

1. Walla Walla *Union Bulletin*, Sept. 27, 1953.

Walla Walla County Gristmills

The flour milling industry in those early times was the main industry for a town and thus the most important one. After these first two mills were built, the Pioneer and the Frontier, other mills followed in rapid succession. As more and more homesteaders came into the region opening up new farms and developing them, more gristmills were needed to take care of the greater quantities of wheat being grown. Gristmills springing up were welcomed by the farmers especially when one was built nearby to lessen the distance to the gristmill. Competition soon could have become a factor, and each gristmill must have sought the farmer's business by trying to offer quick, satisfying service.

The peak for gristmills by the 1870s still had another fifteen or twenty years before the downturn would find the gristmills' usefulness dying out. But now they were in full cry with no one seeming to have the slightest realization of their early doom, for which some gristmills were earmarked as victims sooner than others. One gristmill after another was busy setting down its pine timbers near a running brook, taking over the beaver's watery home as it started its granite stones grinding the farmers' wheat, with no thought of its time limit. In the late 1880s, seven mills were operating within a thirty-mile span in the Touchet Valley near Walla Walla. Tiny towns in other Eastern Washington counties were also building gristmills at a rapid pace, some of them only a few miles from each other.

During the Frontier Mill's building in 1862, and about the same time, another mill was being constructed by E.H.Isaacs, with more mills in the planning stages by other millwrights.

All through the years, Almos H. Reynolds had remained a widower, his first wife and child having lost their lives on their way West, but in 1861, Mr. Reynolds remarried. He took as his wife Mrs. Ransom Clark, a young widow who had brought her small family to Walla Walla to complete the donation claims staked out by her husband but not completed before his death. The Frontier Mill had been built on one of these donation claims.

Reynolds since first coming to Walla Walla, had been active in civic affairs in the town and county and was always interested in the betterment of Walla Walla and ever ready to help in many capacities. He was a capable, level-headed man whose decisions were well thought out and sound. He served several terms on both the city and county commissions of Walla Walla County and was a county commissioner in 1867, resigning in 1869 to be appointed again in 1878 and serving until 1881.

In 1872, around the time McCalley left his employ, Reynolds is said to have sold the mill and after that it changed hands several times. Its existence from then on was frought with troubles for it was twice consumed by fire, to be rebuilt, however, after each blaze. In 1886, Peck and Gilbert are said to have bought the mill owning it and operating it for several years. In the early 1900s the mill came under the ownership of the Walla Walla Mill Company, consisting of Messrs. Coyle and Reynolds. It was completely updated under their management and equipped with the new roller-sifter machinery.

Much later, in 1947, the old Frontier Mill, called the *Coyle Mill*, was used then to grind stock feed only.

Reynolds, during his productive years, had taken on other interests along with his milling, some of which he had been involved in before 1870. He was one of the incorporators and a supporter of the Walla Walla and Columbia River Railroad. In 1869, he and J.H. Day had opened the Reynolds and Day bank, and this bank was said to have been the 'first real bank in Washington Territory.' The Walla Walla Statesman in July 12, 1869, noted, "...and when the vault is completed they will be ready to transact any business in the line of banking and exchange. They will draw bills of exchange on all the leading cities, buy and sell greenbacks, deal in gold dust and generally occupy the position held by Ladd and Tilton in Portland..."

Although in 1878 the bank merged with the First National Bank founded by Levi Ankeny, Reynolds was a member of the board of the First National Bank until his death.

During the years he was building mills in Oregon in the 1850s before coming to Walla Walla, a certificate of deposit—the very first one issued—was given by the Bank of Portland and issued to A.H. Reynolds, Esq. on June 4, 1859, for the sum of $150.

Looking backward to the time when Reynolds may have searched his conscience for the best course for him when his two partners wanted to use the gristmill for less worthy purposes, it must be stoutly apparent that Almos H. Reynolds could never have sanctioned any other enterprise for his gristmill than that of milling flour.

Walla Walla County Gristmills

COYLE'S MILL

Only brief accounts and a sketchy history seem available for the *Coyle's Mill*, but this is remedied easily, for the Coyle's Mill is directly related to the early *Frontier Mill*, a direct ancestor of the Coyle's Mill. Thus the history of the Frontier Mill would seem to be the history of the Coyle's Mill. The Frontier Mill was the second mill built in Walla Walla and was constructed in 1862 near Prospect point on the Yellowhawk Creek by A.H. Reynolds. As the records show, Mr. Reynolds built the first mill, the Pioneer Mill, also in Walla Walla.

Through the years the Frontier Mill changed hands many times. In 1886, Peck and Gilbert bought it, the mill then operating under the name of *Star Mill*. They ran the mill for several years.

The exact date unknown, but possibly around 1900, John D. Coyle bought out Peck's interest and became the other partner of Gilbert. Around 1906 or 1907, a new roller system was installed in the mill, eliminating the granite millstones and increasing the capacity of the mill, renaming it *Walla Walla Flour Mills*.

A Hugh Tinker soon bought out Gilbert's interest becoming a new partner of Coyle. But this arrangement changed again in a short time when Tinker sold his half interest to Martin Campbell. All these changes in ownership and other mill activities as well, came to a sudden halt in 1916 when the old mill caught fire, and burned.

Possibly discouraged, but none the less determined, Coyle immediately rebuilt the mill. Its new exterior shining and the old machinery (used from the burned mill) wiped clean, oiled and working again must still have held appeal and promise, for in 1917 Harry Reynolds bought thc dangling half interest left by Martin Campbell. Reynolds had more than just a passing interest in the mill structure, for it was his grandfather, A.H. Reynolds who had erected it in the first place. Reynolds also may have gone in as partner at this time chiefly to help Coyle financially to keep the mill running under adverse conditions following the fire. Later, Reynolds pulled out, selling back his interest to Coyle, who now became the sole owner. Coyle ran the mill successfully, carrying it on by himself until the year 1921, when his part in the operation was brought to a sad, abrupt end by his death .

Razing of the manufacturing portion of the old Walla Walla Flour Mill on Yellowhawk Creek near Prospect Point. Coyle's Mill, the early *Frontier* on this site was built by A.H. Reynolds over 100 years ago. Section being removed is in the foreground. The storage section is the other portion of the mill.

Courtesy Whitman College, Penrose Mem. Library

Coyle's son, John D. Coyle, now took over the operation of the mill after his father's death. He ran the mill for several years but he also died in the 1920s. John's widow, Mrs. Elizabeth Coyle assumed management, handling all the business affairs until her son, Walter, was old enough to take over. According to Walter Coyle's report, the mill ceased manufacturing flour in the 1920s and was converted over to the production of livestock feeds and cereals.

Mrs. Coyle, still owner in 1938, eventually sold the mill to John Yenney who took it over and ran it. In 1962, Yenney had parts of the old mill dismantled and torn down, especially the "manufacturing part of the mill," retaining the grain storage portion of the mill .

Walla Walla County Gristmills

EUREKA MILLS Originally, *Agate Mills*

The exact date of the construction of the *Eureka Mill* is not known but thought to have been in the early 1870s. It was built by two millwrights by the names of Ritz and Schnebly, and located "about one-quarter mile below the McCalley Mill," the McCalley mill being built a mile west of town. A man by the name of W.C. Painter is said to have operated the mill as its first miller. At first the mill was known as the *Agate Mill* but it was soon changed to *Eureka*.

The Eureka Mill was a large and an important one from the beginning. It had three run of granite millstone and was operated by a water-powered turbine. It had bolting equipment for sifting the flour into finer, whiter grades, the bran and shorts a by-product, and there was cleaning equipment. There was a well-planned system of conveyor belts for moving wheat or flour to other floors to storage bins or hoppers.

It is remarkable at that time, although there were gristmills dotting nearly every stream of water, competition among them did not lessen any of their efforts or diminish their owners' enthusiasm for greater production. Even as early as 1869, there were already six strong gristmills in Walla Walla County with a seventh one planned, the list growing rapidly all through the 1870s and 1880s. The Walla Walla Weekly Statesman, September 3, 1869, states: "Mr. Chapman, a practical miller, has lately bought a lot of land about a half mile below town, on a line with Mill Creek, and proposes, as we are informed, to at once erect a flouring mill....We already have six mills in the county, and it is questioned by some whether a seventh will prove a paying institution. We give a list of the mills in the county in the order in which they were built: Pioneer Mill, Frontier Mill, Excelsior Mill, Wait's Mill on the Touchet, Reese's Standard Mill, and the Milton Mill. The coming fall and winter, should Mr. Chapman adhere to his determination, will witness the erection of the seventh mill, which we understand is to be smaller than Reese's and less expensive in construction."

There is no record that Mr. Chapman built his gristmill, but an intensive building of gristmills took hold and continued on even into the 1890s. By then, the new technique of rollers was coming into use and most new mills were fitted with this equipment. The older mills were nearly all enlarging and changing to the new roller system of producing flour.

The *Eureka Mills* when first built in the early 1870s.
Courtesy Whitman College, Penrose Mem. Library

Those mills that did not change to the new stayed in business only a short time, and even those who installed the roller equipment when they built the gristmill eventually had to give way, for the flour making industry was widening its horizons and was on the threshold of a tremendous development which did not include the flour mills in small towns.

The first owners of the Eureka Mill sold it in 1880 to two men by the names of Welch and Schwabachner. They operated it a short time then sold out to two energetic, business-minded as well as civic-minded brothers by the names of Frank and Fred Dement. These two men took over the mill's operation, carrying on the 'Custom' trade for the farmers and taking on a large 'Merchant' trade, using their trade name of *Dement Brothers* coupled with the old firm name, *Eureka Mills*.

A few years later, in 1887, the two brothers realizing they must meet the standards set by the newer mills who had either installed the new roller equipment, or had been built with this machinery already installed, now decided to convert to this new method of producing flour. They began rebuilding the mill at great expense, enlarging the structure, adding a large loading and storage structure on the left side of the building. They took out the three run of granite millstones and installed a complete set of steel roller-mills, compactly contained in neat-looking cabinets. They also put in other equipment and machinery to go with the rollers: automatic sifter, cleaning apparatus, bleaching equipment, chain conveyors and other cupped belts, chutes, etc., making the Eureka Mills one of the best equipped and finest in the country. The rollers came from Todd and Stanley of St. Louis, Missouri, "the best money could buy." The production of the mill was 225 barrels of flour every twenty-four hours stepped up drastically from the old turnout of 100 barrels a day.

Now with the mill's updating, new orders increased accordingly and demands for its flour came from all parts of the country and from foreign regions. The mill operated thus for ten years at which time the brothers then decided to add another story to the main part of the building. At this time, needing extra power, they put in an immense underground barrel-flume from Mill Creek down to the mill which was a distance of 1,050 feet. They still wanted water power to run this mill although they were now hooked up to electricity to augment the power from the creek.

The *Eureka Mills* after being remodeled and enlarged in 1887.
Courtesy Whitman College, Penrose Mem. Library

Fred G. Dement

(Courtesy Whitman College, Penrose, Mem. Library)

In addition to their milling, the brothers bought and sold grain and gained a wide reputation for honest and "square" dealing throughout the valley.

Their brands of flour were known as "White Spray," a patent flour; "Dements Best," straight or unpatented; "Walla Walla," clear, and "Excelsior," a superfine grade. These were in such demand that it was often difficult to keep up with the constantly increasing orders, large shipments going to all the states as well as to Alaska, British Columbia, China, Japan and South Africa.

The Dement brothers were outstanding men, both as business men and community-minded workers in the town and region of Walla Walla. They lent their names and their time, and often their financial support to many worth-while community projects and could be counted on to give their help in any way needed.

Walla Walla County Gristmills

Frank S. Dement

(Courtesy Whitman College, Penrose Mem. Library)

Frank S. Dement was born November 3, 1853, in Oregon and learned the printing trade at an early age. He purchased and ran the Oregon City *Enterprise* and was also treasurer of Clackamas County in Oregon for two years. His health failing in 1880, he came to Walla Walla to enter the grocery store business. A year later he was joined by his brother, Fred G. Dement, two years younger, and they both decided to 'engage in the milling trade.' Fred, like his brother was interested in civic affairs and also an untiring worker in the Republican political party.

Fred G. Dement was born September 3, 1855, at Oregon City, Oregon. The family moved to California and Fred attended the Heald's Business College in San Francisco when he was a young man. Coming to Walla Walla in 1880 he shared the grocery business with his brother before they both went into the milling business. Fred was long identified with the republican party and was a delegate to the Everett convention. He handled the superintendency of the mill, assuming the management of the business of the firm.

Vanished Gristmills

In those early days, farmers did not have the help and guidance from institutions, such a colleges, in carrying on special work to improve grades of wheat for growing in certain regions. Farmers grew wheat then which had probably been grown in other sections of the country, such as the eastern part of the United States and much of it was not suitable to the western climates. An interesting comment on early brands of wheat grown in southeastern Washington shows the trials and efforts constantly expended for better wheat crops.[1]

In the 1800s, *Yellow Club* wheat was grown in this area. This wheat was more suited to the foothills than to the drier Snake River edge of the county. The late George Welch summed it up with these words: "*Club* wheat matured late, the crop often being caught by hot north winds and we hauled pinched wheat down the gulches on rough roads, sometimes of a value of twenty-five cents a bushel. The Flat was broke. We heard of a wheat that matured earlier. We bought and planted it. It was called *Bluestem*, and was suited to our country.

"The Dement firm imported this wheat from Australia through William Dunbar, an importer located in Portland. The grain arrived in four-bushel jute coffee sacks. Among other purchasers, George Delaney bought one sack. The resulting crop was purchased by Alexander and William Reed and peddled over the country. This high-pressure salesmanship established the wheat in this area."

1. Walla Walla, *Union Bulletin*, Nov. 15, 1953.

Frank Dement, when a young man, had been advised by his doctor, "If you want to live, go up in Eastern Oregon—go up in the bunch-grass country," for at that time he had come here with a respiratory illness. But he was to live for another sixty-four years,—a total of ninety-four years, his age at the time of his death on September 8, 1947.

The Dement brothers ran their mill for thirty-five years, finally selling out in 1918 to the *Northern Flour Mills Company*. The Eureka mills was completely destroyed by fire in the 1930s and never rebuilt.

Vanished Gristmills

WALLA WALLA COUNTY GRISTMILLS

WAIT'S MILL

Also: *Washington Mills*
Preston Parton Milling Co.
Preston-Shaffer Milling Co.

Nearly every early millwright and miller who contributed so much towards the development of the West's first industry, flour milling, came from elsewhere. Some were from the eastern part of the United States; Iowa, Illinois, Kansas, Missouri, etc., where they had learned the trade, or from a foreign country—Germany, Norway, or other places. Likewise, the farmer who became his neighbor was usually from another region, with more or less the same regional background as the millwright and miller.

Sylvester M. Wait, a well-trained millwright and miller was from Waitsfield, Vermont, a town named after Sylvester's grandfather, General Benjamin Wait. Sylvester as a young man, joined others coming West. Before he came to the region which would bear his name, Sylvester was at Rogue River, Oregon, operating a gristmill there. When the Rogue River Indian War broke out this business venture came to a halt, and Wait moved up to Lewiston, Idaho, to run a dairy farm until he could get back into his field of milling again.

Dennis Willard, called "Doc," was perhaps the first farmer to come to the then unnamed Waitsburg region in the Touchet Valley near Walla Walla. He originally was from New York, starting West as a young man and reaching California in 1850. He was in the mines there for a while, fell ill, and after recovering went on to Oregon to settle in Benton County. Still looking for a more suitable location he moved on up to the Touchet Valley in 1859 and satisfied with this country, claimed for himself 160 acres of land. The land took on the characteristics of a farm after he had grappled with pine tree and sagebrush from sun-up to dark and had followed his team and wooden plow around rough fields to make the sod ready for a crop of wheat. A few other folks were doing the same thing, clearing their lands and raising the initial wheat crops of the region, which from the first were surprisingly good.

For some time the few farmers in this region had been talking 'gristmill.' They wanted one badly for they had to drive their teams and wheat all the way to Walla Walla to Reynolds' gristmill on the Yellowhawk to have it ground into flour. They often had to remain many extra hours taking their turn, for the mill was a busy one with lots of farmers bringing their wheat there.

Walla Walla County Gristmills

It was at this point that destiny, or luck, or fate may have stepped in to bring the right millwright, Sylvester M. Wait, to that Waitsburg region by way of a chance meeting with the right farmer, "Doc" Willard. At the moment of their meeting, Doc was "freighting" goods into the Lewiston area on his way to the mining settlements in Idaho. Wait was in Lewiston for some business reason connected with his dairy farm near by. After their first handshake, the subject of mills came up, Doc perhaps asking,

"What might be your line of work?"

"Milling. Been a miller all my life. Millwright, too. Still'd be in that work down on the Rogue if I hadn't joined up with Company 4. Indian War, you know. Bad one. Right now—dairying with a few cows up near here. But milling is my line and I'm mighty anxious to get back into it."

Sylvester M. Wait, Millwright and Miller.

(Courtesy *The Times*, Waitsburg, Washington)

Vanished Gristmills

Only a few facts later found both men vitally interested in the mill subject with certain developments already in fast progress. Sylvester M. Wait was extremely impressed to know 5,000 bushels of wheat were being produced around Mr. Willard's region, and that folks there badly needed and wanted a gristmill. Also, that there was plenty of water for power. Doc even assured Wait he would donate a millsite if Wait would start a mill. Before the two men parted, Wait promised he'd be down in Doc's section in about a month, just long enough time for him to dispose of his dairy holdings.

By the time Mr. Wait arrived, Willard had already contacted the farmers telling them of this excellent opportunity for a gristmill and a man to take charge. "He's a good man," Doc would have said. "And he can handle it. Now, it's up to us farmers to stick together and support this mill."

When Sylvester Wait met with them and they'd discussed it thoroughly, agreements were soon down on paper and signed with the farmers promising to hold their grain until spring at which time they'd sell to Mr. Wait at the mill for $1.50 a bushel. Wait couldn't see how he could lose on this with flour worth $14 a barrel. Willard and another neighbor, William Bruce, more than did their part, with Willard donating ten acres of ground for the millsite and giving the right-of-way on his land for the millrace. Bruce helped raise $14,000, part of it credit, for the construction of the mill, although Wait himself had $1500 or more of his own to put into it.

The gristmill got underway by 1863. Taking a week or ten days off from supervising the work, Wait went to San Francisco and put in an order, on credit, for all the machinery, returning to find the building taking shape fast. A lot of hard work went into its construction. Huge logs for the main beams were cut and brought from the Jasper property in the Blue Mountains. Andy and William Jasper hauled them the long distance to the building site. There they were hand-sawed, and cut and trimmed with the adze, the dependable handcutting tool with its curved blade, the men working the beams into the proper sizes and shapes for the building's supports. But lumber otherwise was so scarce and high-priced every available plank and stick of scrap lumber was used if in good shape. It is said, "An old sheep corral and other structures were bought at high prices in order to get boards enough to enclose the mill."

When the mill was finished in 1865, the building was a large, four-story structure with a good-sized overshot water wheel. This took water into its cups from the top of the wheel, the water brought along a wooden box flume. The flume was connected to the mill ditch, or millrace, which came from a dam with water from the Touchet River, a clear, clean running body of water with unlimited power resources for the farmers' uses. There was one run of granite millstones, each of the two stones thirty-six inches in diameter, with the mill's capacity 50 barrels of flour in a day. The bolting equipment was a drum-like device using centrifugal force. There was a screening frame attached to it which sifted out the bran and the shorts, the two or three screens fitted with silk material of different sized meshes.

The dimensions of the four-story building were 30 X 50 feet, built purposedly large to handle all the wheat expected without the need for more space for many years to come; or at least that was their thinking at the time. The building, painstakingly and skillfully constructed in the manner of the millwrights of that time, was built with mortised joints, and instead of nails, secured with pegs or pins made of wood from black walnut trees.

In those early times, farmers around Waitsburg are said to have been cultivating about 1,800 acres of wheat, and from the first moment the mill opened up for business, farmers flocked to it from everywhere with their wheat.

The mill ran day and night. In order for the miller on duty during the night, Mr. Beattle—and his assistant, to see during those dark hours, candles were set around on the sills or at other safe places and lighted for their use. Their tiny, flickering half-lights may have been better than total darkness, but little more so although it let them work through the night time. The Jasper brothers' sister, Lois, later Mrs. A.G. Lloyd, spent many hours daily making candles for the mill, dipping one after the other into the deep pot of hot wax until each had grown large enough to stand lighted in a candle holder. Their quavering beams must have turned the flour-dusty interior into a strange theatre of eerie, shadowy dancers, but perhaps Mr. Beattle and his helper were too busy to notice.

In those years around 1865, working by candle light was common practice. The physically strenuous labors were done outside usually in the daytime, with a farmer and his wife saving the lesser active chores for the evening hours by candle light. This might be churning, or the finishing up of meat preparation after the butchering, shelling the corn or working with other grains, and any number of other farm duties. All life centered around the important wheat crops; preparation of soil, planting, harvesting, then turning the grain into food products, such as flour.

One hundred years in the future from Wait's time, in the 1970s and beyond, machines everywhere would make work easier, especially on farms. It is difficult now to conceive of fields of wheat in the middle 1800s sown by hand, seed thrown out over fields with wide circles of the arm. Later, when ripe, the grain cut with that same wide sweep of arm and hand, swinging the cradle scythe. Next, there was the flailing, or separating the wheat from the chaff, and the winnowing to further clean it, the winds helping blow out the unwanted dirt and pieces of straw. By the time the wheat kernels had been sacked up in homespun sacks and taken to the miller, if there was a mill somewhere close, the farmer's hardest work with his wheat crop had been accomplished. Having a gristmill in their own community was a distinct blessing, for it saved the farmers a long trip to a mill in another town, or from having to grind the wheat into flour themselves by some difficult and laborious home method.

Names of some of the farmers in the Waitsburg area bringing wheat to the mill, might have been J.H. Morrow, Wm. McKinney, Walters, Sherm Pettijohn, Andy and Wm. Jasper, Albert Lloyd, Dennis Willard, Wm. Bruce, A.C. Dickinson, J.W. McGhee, Wm. Whitaker, W.J. Billups, Henry Livingstone, Anderson Cox, Wilson, Purrington, the Rayburns, Levi Dickeson, Lewis Neace, W.P. Bruce, David Gerking, Elmer Stonecipher, George Loundagin, T.J. Hollowell, H. Preston, C.W. Wheeler, E.L. Powell, George Rice. There were many, many others living there then who used the mill.

The years of the 1860s in southeastern Washington's history, found everything happening at once. New settlers, new wheat crops, new towns, new gristmills and new millers to run them, and a new and different breed of men, taut-figured creatures with a gold tinge to their eyes, feet with worn shoes trodding paths through the Walla Walla region to the mines in Idaho. The path stopped at Walla Walla and Waitsburg, men's feet and horses' hooves resting a while, as bags were filled with supplies. For the most part supplies were represented as slabs of bacon and sacks of flour. Many a man who had not been able to wheedle a Walla Walla merchant into giving him flour on credit, would stop at Wait's Mill, having somehow secured a little money somewhere between the two towns, and buy a sack or two here. The word "cash" had come to mean just what it spelled out and if they didn't have the 'hard money' there'd be no flour for them. But the cash that came in, became a welcome credit item in the Wait's Mill's ledger, for it helped balance out the debt column of red-inked figures, capital necessary to get the mill under way. But Wait was an astute business man, and the mill seemed to be a steady money-maker.

When Sylvester Wait came to Walla Walla to begin the mill, he brought his wife and good-sized family of five sons and three daughters. Wait was born in Waitsfield, Vermont, on May 20, 1822, and he and his wife, the former Mary Hargrove of Illinois, never regretted coming West. With his sound training in the flour milling trade, Wait had always found opportunities for work, and had always been in demand for millwrighting or as an operator of a mill.

A kindly man, Wait gave much thought and weight to his decisions, and took into consideration another person's needs and purposes along with his own when any important venture was in progress. Straight-forward in his manner, open and honest Wait had the respect and trust of the farmers who brought their grist to his mill.

Wait took an active part in civic matters and helped in the community's growth. He was county commissioner in 1867. In 1870, he built an important mill at Dayton which helped in the growth and development of that town. He had other successful business ventures, such as merchandising businesses, and he was a successful promoter of other enterprises, including railroads into that area. He founded the settlement of Elberton, developed it and ran a gristmill there, and he was paid homage in 1868 by the Waitsburg people who named the town after him.

Before the area's vote settled on the name of Waitsburg, the fledgling town underwent other appellations. At the beginning, it was simply called *Wait's Mill*, people sometimes affectionately referring to the town as *The Burg*. Other names were tried; *Horsehead City, Waitsburgh, Delta Mills,* and *Delta*. Delta wasn't too far-fetched since the mill was built on the delta formed by the confluence of the Touchet River and Coppei Creek. But after W.P. Bruce unselfishly gave up his farm to civic needs, platting it in 1868 and turning his property into townsite lots, the town was then officially named, *Waitsburg*.

By 1870, only five years after it was built, the Wait's Mill was already becoming crowded and Mr. Wait realized he must either enlarge the building, or sell to someone else and possibly start up another new gristmill in another location. He would have had no hesitation at all in beginning another mill in a new place for this brisk industry seemed at its height, with no let-up of productiveness in sight, and with competition little matter for concern since there was plenty of business for everyone. One gristmill after another was being built throughout the region, the area becoming a mecca of them, their flour dust a symbol of their productivity, their rushing streams an endless proof of their longevity and endurance. If some far-sighted visionary may have hinted of the possibility of a sudden halting in the use of water-powered gristmills and their revolving granite millstones, his words were either not forthcoming or if spoken, unheeded.

In 1870, two brothers by the names of William and Platt Preston, made Mr. Wait an offer for his mill. Wait did not sell out entirely giving the brothers a half interest. At this time, a store was opened in a portion of the mill building. Before long, however, Wait sold the remaining interest to them, no longer holding any interest in the mill he had started only a few years before.

Under the Preston brothers, the name of the mill was changed to *Washington Mills*, and Wait, now forty-eight years of age started building a mill farther up the Touchet River at a stage stop run by Jesse Day, this location later called Dayton.

The Preston brothers immediately took on the task of enlarging and expanding the mill, for under their new plans and management the mill business was about to skyrocket. They had secured a monopoly on the flour needs of the mining camps,

From an old print of the Preston-Shaffer Mill made about 1900

(Courtesy, *The Times*, Waitsburg, Washington)

becoming the only mill in that region to supply them with this necessity. This new business greatly increased the mill's production, already large, and also the labor force, for men with teams and wagons were now needed to haul the big loads of flour to the mining camp towns.

A larger overshot water wheel was installed and the flume was raised to a greater height to accomodate it. With this greater amount of power, four new sets, or run, of millstones were installed. These new millstones were much larger than the first 36-inch stones, the new ones weighing from one thousand pounds to twenty-eight hundred pounds and around four or five feet or more in diameter.

Walla Walla County Gristmills

The new millstones came from France, "by the way of England, then to New York where they were expertly fashioned at the Broadway Foundry, shipped around Cape Horn by clipper ship to Portland, Oregon, then up the Columbia River by boat to Wallula and hauled overland to Waitsburg by ox-team."

In 1871, the Washington Mills took in a third partner, an outstanding miller from Albany, Oregon, by the name of Frank Parton. The mill's name was then changed to *Preston-Parton Milling Co.*, and at this time the firm became incorporated for the first time.

From the 1880s on, flour milling underwent a great change. Burr mills, as gristmills were often called, were becoming a thing of the past for a newer method of milling was taking hold throughout the country with all mills who wanted to survive and continue producing longer, converting to the new. This was the roller system. This system was developed by a mechanical genuis, G.A. Bockler in 1662, but it did not take hold until much later, not being used in the East until the early 1870s and in the West a short time later.

Rollers offered several advantages, the main one being the more uniform and gradual breaking up of the wheat kernel automatically to finer and finer grades of flour until the desired product resulted. It also did away with the tricky use of the revolving granite millstones, which had to be dressed often to keep them sharp for proper grinding. Often, however, many mills converting to these compact roller-cabinets usually kept one run of the old millstones to be used for special orders.

The Preston-Parton Milling Company now decided to convert to this new method of milling by the use of rollers, and in 1881, they took in a new member of the firm, William B. Shaffer, who was well versed in milling techniques having been associated with the great milling industries of Minnesota. The firm changed its name to *Preston-Shaffer Milling Company*, the company now enlarging the mill and installing six of these roller machines. They removed three-run of the four-run of millstones, but kept one set of them. The power was greatly stepped up with the addition of a turbine, an "Eclipse Turbine," operated by water from the flume and with the alteration of it for a standpipe to bring the water to the turbine.

The output of the mill by the 1900s was 400 barrels a day. The mill itself was an imposing, five-story structure, with its additions and wings and four large warehouses and the railroad track running parallel to it.

As time continued, more and more improvements came to the mill, wheat being brought into it through an underground tunnel as well as by way of an overhead conveyance. But electricity was not installed in the mill until 1936, when water after that time was no longer used. Unhappily, the once clear Touchet River

One run of smaller millstones kept in the mill after converting to rollers. These millstones were removed in 1925 and placed in the city park.

(Werner Lenggenhager photo, Courtesy Spokane Public Library)

had become sluggish and turbid, the water no longer useable because it contained so great an amount of sediment which filters could not clear well enough for the turbine's needs.

In 1915, Mr. E.H. Leonard joined the organization, to become president in the later years of the mill's operation. In these productive years, the capacity of the mill increased to 750 barrels a day, operating on a twenty-four hour basis the year around, with the mill employing twenty men.

The mill ran a number of years longer, an important flour milling company always and a boon to the town of Waitsburg and all that area, but it finally ceased operating to close its doors as a mill in 1957. By this time, and even before, (although this mill had remained operating long after other mills in that vicinity ceased), milling had become an operation taken over by the large corporations, and was thus no longer a profitable business for individually-owned firms such as the one in Waitsburg.

Walla Walla County Gristmills

Early Picture Wait's Mill at Waitsburg

The history of the Waitsburg flour mill is a success-story from its beginning in 1865 to its closing in 1957. At no point in all its years of activity did it ever falter from lack of business. Its products were of the highest quality, sought after throughout the United States and foreign countries as well. Its standards were high and its owners and its millers were outstanding men of sterling worth and ability. The Wait's Mill set an example of excellence for all other mills in that part of the country.

It long outlasted its first originators. Its builder and first operator, Sylvester M. Wait died at the age of 69, December 28, 1891, and was buried at Dayton. It outlived one of its staunchest supporters and instigators, the first farmer of that region, Dennis "Doc" Willard, who died near the age of 63 in 1893.

The old mill building still remains, intact and sturdy, serving in another capacity with wheat, the Touchet Grain Growers Company occupying offices in the structure.* The machinery is still in the building and evidences of its past labors are still apparent in the maze of drive shafts, belted wheels, pulleys and intricate equipment. The big mechanical sifters as well as the rest of the machinery look as if they need only turning on to put them in operation again.

If the faint scent of flour dust sifts through infinitesimal cracks in a wall built long ago, now housing an office, no one will mind. This was a town where white flour dust mingled with black rich soil of the region to help build that town. Here wheat was the first crop grown. Most of this wheat raised then and in the years following, was made into superior grades of flour in the Wait's Mill, and this mill operated for ninety-two colorful, productive years.

*For preservation purposes, listed in National Register of Historic Places.

Vanished Gristmills

WALLA WALLA COUNTY GRISTMILLS

NORTH PACIFIC FLOUR MILLS — First called, *Excelsior*

H.P. Isaacs, one of Walla Walla's outstanding millwrights and millers, on August 3. 1862 began building a gristmill on Mill Creek just east of Walla Walla. This location has in later years been designated as "at the present intersection of Boyer avenue and Division street, on what is now Wildwood Park."

The *Excelsior Mill* was built about the same time as the second mill built, the *Frontier Mill* on the Yellowhawk Creek constructed by A.H. Reyholds in 1862. Reynolds, also built the first mill, the *Pioneer*, dissolving his partnership in 1862 with the other two men connected with this mill and immediately starting to build the *Frontier Mill.*

When Isaacs began the Excelsior mills, he had a partner by the name of Metzger, an 'old timer' in the millwrighting business, skilled in all fields of carpentering and millwrighting. He worked with Isaacs in the construction of the mill, going "fifty-fifty" on the deal, each man taking half the responsibility of building and financing. Later, however, when the mill did not produce enough to bring Isaacs a profit, he bought out Metzger and operated the mill himself.

The mill at first was mainly constructed by Isaacs for the purpose of supplying flour for a store which he owned and ran in Boise City. The Boise store was a thriving market for it was close to the mining areas which were experiencing the 'first throes of gold excitement.' Isaacs had no difficulty in finding men who would freight the flour from Walla Walla to Boise at the rate of "two bits a pound."

H.P. Isaacs was a well-known and well thought of millwright and miller in the 1850s and the 1900s, which might be said to have been the time span encompassing 'hey-day to swan-song' for gristmills in Eastern Washington. Isaacs could well be called one of the very first early "chain- operators," for he built a number of mills in Washington and Oregon, operating them under the name North Pacific Flour Mills, each under his control. Besides the Walla Walla mill, built in 1862, he also built and established one at Boise in 1864; Middletown, Idaho, 1865; Prescott, 1883; Wasco, Oregon, 1899, and one early mill which he is said to have established at the military Fort Colville in 1858.

Born in Philadelphia in 1822, Isaacs started west before he was out of his teens. Before coming to Walla Walla he operated a store in The Dalles, Oregon, also

H.P. Isaacs

Courtesy Whitman College, Penrose Mem. Library

owning at one time the historic Umatilla House there. Later, as one of the outstanding civic-minded citizens of Walla Walla, he fathered the bill for the 1886 territorial legislature which established the state penitentiary in Walla Walla. His very first business venture in Walla Walla was to operate a general store.

NORTH PACIFIC MILLS (ISAAC'S MILL) WALLA WALLA

Courtesy Penrose Mem. Library, Whitman College

Isaacs may have gotten into the millwrighting and flour milling business in Walla Walla a little sooner than he intended because of an incident precipating this venture.

He often made periodical visits to Philadelphia to buy goods and stocks for his store, and one time before leaving on one of these trips he was approached by two men asking a favor. These men were anxious to build a gristmill and get into the milling business, a sure-fire lucrative enterprise in those early days, or so it sometimes seemed to the novice. They asked him to buy machinery for them on this trip East, so they could set up a gristmill and get started in the business.

Isaacs did this, and the machinery for them was shipped 'around the horn' and up to Wallula, but by the time the machinery had arrived the two men, lured by the richer promises of the nearby mines, had left Walla Walla. Isaacs, now with the machinery and equipment for a gristmill on his hands, concluded he had best use it himself to build a flour mill at Walla Walla and this was the **Excelsior,** or *North Pacific Mills*. Metzger, his partner then, installed the machinery and equipment in the mill. There was one run of millstones, and these "stones" were quite unusual, for instead of being composed of the usual granite quartsite or of sandstone, these were fashioned out of iron. It is likely a turbine was used in this mill, indicated by the output of 75 barrels a day, one run of millstones run by an overshot wheel usually capable of turning out only 50 barrels a day. The water to run the turbine came from Mill Creek collected in a pond to supply a steady head of water.

A short time after Isaacs took over the sole ownership of the mill, he updated it, removing the iron burrs, or millstones, and changing the mill to roller equipment. His mill is said to have been the first in the state to convert to rollers and this may have taken place in the early 1870s, exact date not known. It was at this time he changed the name of the mill to *North Pacific Mills*. The pond that supplied the water to run the mill is saidto have been situated at the present corner of Evergreen and Roosevelt streets in Walla Walla.

The old pond, like most mill ponds and water ways, was an attraction winter or summer for youngsters. It was in a charming park-like setting with pines and fir trees bordering its edges. People called it Aiele's Pond and the location was thought to be about the same as that of Pioneer Park now.

From the mill pond, a wooden box flume ran diagonally to the corner of Boyer and Division and then to the mill. At a point on Boyer and Division, a large iron grating was placed in the ditch to catch debris, weeds and foreign matter collected there which was removed periodically so it would not seep through and clog the mill's machinery. But the iron grate played a sad part in a local tragedy which occurred at one time in the 1880s.

A small boy, son of Mr. and Mrs. McGrew, playing in the water of the flume, fell in and was carried down by the swift water until he came to rest against the metal bars of the grate. A doctor living close by, Dr. Chase, was quickly called after the child was discovered missing only a short time after the accident, but it was too late for the little boy had already drowned.

At one time in 1872, Andy McCalley operated the North Pacific Mills as a miller for Isaacs before he purchased a mill of his own west of Walla Walla.

In 1878, Louis Scholl ran the mill and continued doing so until 1883, then taking a similar post in Isaacs' new mill at Prescott, to remain there for thirteen years.

The Scholl brothers, Carl and Bismark worked for Isaacs as millers with Carl Scholl operating the mill until 1898, at which time he and his father and brother went into business for themselves.

One day Carl Scholl came to work and found all the power gone, the mill quiet and motionless. After turning the valve to close off the water, he began checking all the gears and belts finding them seemingly in good working order. He went on to the location of the turbine to examine it and there he found the trouble. A dead man was lying on top of the turbine in such a position that the gearing was obstructed. No one ever seemed to be able to solve the mystery of the unfortunate man found there and it could only be assumed as time went on and no information ever came to light to clear the situation, that he must have committed suicide, or died of a heart attack.

During the years of operation, the mill was often improved and additions built on to the mill to hold new equipment at different times. It is believed the North Pacific Mills was the first on the Pacific coast to install rollers, and as has been mentioned, this was thought to have been done in the early 1870s. Such a development made this mill one of the most important mills in southeastern Washington.

The mill building itself was large, four stories high, the capacity of the mill growing from 75 barrels using the old iron millstones, to over 250 barrels, and some historical accounts claim, 500 barrels after the roller equipment was installed. Its products enjoyed a wide market at home and abroad.

Isaacs never sold this mill and large grain storage house beside it. This remained under his ownership and control until the year 1899, at which time he closed it down. A number of other mills were operating throughout the Walla Walla region by then, and Isaacs felt the mill could not compete with others unless it were renovated and expanded beyond its present production and capacity. But although the North Pacific Mills as a mill was finished, H.P. Isaacs its owner, was not. He was busy constructing a new mill at Wasco, Oregon, one in which he planned to use steam power. Isaacs took out some of the machinery of the North Pacific Mills and sent it to Oregon. He had other parts of it sent to Prescott to be added to the large mill, North Pacific Roller Mill, there he had constructed in 1883.

In its final days, the North Pacific millsite and the water rights were sold to the Walla Walla city water works and the North Pacific Mills became relegated to the flour mills of the past; one of the first excellent and extraordinary ones to vanish from the scene of early flour milling, but by no means the last.

Walla Walla County Gristmills

PRESCOTT MILL,
NORTH PACIFIC ROLLER MILLS, B

One of the most successful of the many milling enterprises of H.P. Isaacs was the large flour mill built at Prescott in 1883. This became the largest mill in Eastern Washington, with a building of five stories.

Roller equipment and a turbine were installed in this mill at the outset, water for power coming from the Touchet River. The mill was well equipped with other up to date machinery—cleaning apparatus and the sifting machinery to go with the rollers.

The Prescott North Pacific Roller Mills could grind five hundred barrels of flour per day, the first mill ever to produce that quantity of flour in twenty-four hours. A barrel of flour weighed 195 pounds, which represented a capacity of 181,000 barrels per year for the mill.

North Pacific Flour Mill at Prescott, built by H.P. Isaacs in 1883.
Courtesy Penrose Memorial Library, Whitman College.

Vanished Gristmills

Roller mill built by H.P. Isaacs at Prescott in 1883 as it looked when first built. Note fuel burning steam engine at right, attached to tender and box car of O.R.&N. Co. Isaacs built a North Pacific Flour Mill in Walla Walla in 1862. Courtesy Penrose Mem. Library, Whitman College

At Prescott, Isaacs bought large quantities of grain from the farmers and built ample store houses for holding the grain for his needs.

At one time Gus Steinke was its miller. Louis Scholl became bookkeeper for the company. In 1881 he surveyed the canal and made plans for the mill at Prescott. where he served as superintendent until 1894. He also in the period 1856 to 1859, drew up the plans for Fort Walla Walla. He was born November 4, 1829, in Baden, Germany, coming to the U.S. when nineteen years old.

Louis Scholl's sons, Carl and Bismark, worked at the Prescott Mill for several years. Carl Scholl lost a hand in the rollers of the Prescott Mill. Milling had its hazards. Revolving gears and large cogged wheels, or belts could catch a miller off guard in a fleeting moment and knock him off balance. A miller never for a second could forget the machinery's capabilities—or its heartlessness.

In the 1880s, Prescott was an important early stage stop, and that business operation, along with the flour mill coming on in 1883, helped the town to grow and develop rapidly. As soon as the Prescott mill was built, a steady stream of farmers brought their grain to the home mill, no longer having to take it the long distance to the Wait's Mill at Waitsburg.

H.P. Isaacs operated the Prescott Mill until 1900, then leased it to the Portland Flouring Mills Co. In 1902, the Isaacs family sold the mill to Preston-Shaffer Milling Co., and in 1904, that firm sold the Prescott Mill to the Portland Flouring Mills Co. They operated the mill until it was destroyed by fire July 22, 1922. It was not rebuilt.

Outstanding men shaped the beginnings of the flour milling industry in Walla Walla County: Reynolds, Dement Brothers, H.P. Isaacs, Metzger, Scholl Brothers, McCalley, Mulkey, Sylvester Wait, the Long brothers and many others. In shaping the industry, these men also helped shape the town of Walla Walla, and the county.

Vanished Gristmills

WALLA WALLA COUNTY GRISTMILLS

CITY MILLS

The two Scholl brothers, Carl and Bismark, and their father, Louis, were with Isaacs at his North Pacific Mills for a good many years. In 1898, they left his employ to start a mill of their own. This was the year before Mr. Isaacs closed down his North Pacific mills. The Scholls pooled their finances and built a flour mill on Palouse street in Walla Walla on the south side of Mill Creek, and called this mill, *City Mills*.

The mill had one run of granite millstone and turned out 75 barrels of flour a day. There may have been bolting apparatus to sift the flour to a finer grade, and with this capacity the mill was undoubtedly powered by a turbine using the water of the Mill Creek.

To channel the water from the creek to the mill, the brothers put in a barrel flume which they buried under ground. This came down what later became Alder street in Walla Walla.

Mr. Allen Mulkey, originally from Pomeroy, Washington, and born in 1887, (brother-in-law of Ray Munns and cousin of Mrs. Ellis Johnson), is said to have had his first milling experience at City Mills, learning the milling trade from the Scholl brothers. He went to work for them in 1909.

Mr. Mulkey's career as a miller is of note. He worked in mills in different places around the Northwest after being at City Mills for many years. He was head miller for Collins Flour Mills in Pendleton for more than twelve years and also with a Portland, Oregon, firm at one time.

It was at the Portland mill where Mulkey devised a process of making a flaked wheat cereal. He patented it and then sold it to General Mills in 1922. Today this is the cereal that is called "Breakfast of Champions."

The old barrel flume which was buried under ground for the City Mills to bring the water from the creek, caved in finally during the 1950s. The street gave way at one point along the barrel flume run. This happened long after the mill had closed down and when the flume was no longer used. Instead of causing consternation or alarm, it brought only a smile of nostalgic appreciation of the long ago methods used in early day flour milling.

Walla Walla County Gristmills

McCALLEY MILL

In 1866, I.T. Reese and a Mr. Kyger who was a well-known merchant, leased the water power and a site from E.H. Barron just below town on Mill Creek. On this site, the two men erected a gristmill, it said to be "one-fourth mile above the Eureka Mill," which was west of town.

The mill, first called *Reese's Standard Mill*, was "the largest and most valuable mill in the county"[1] being equipped with three-run of stone, each four feet in diameter. There was a bolting machine to sift the flour to finer grades.

The mill was purchased in 1872 by Andrew McCalley, considered at that time one of the most capable millers in Walla Walla county. Unfortunately, soon after McCalley's purchase of the mill, it burned down. Using the machinery and equipment from the old mill, McCalley went to work and rebuilt the mill running it himself for fifteen or twenty years.

After McCalley's death in 1891, the mill was taken over by his sons, who shortly afterwards sold it to W.H. Gilbert. Gilbert at one time, 1886, had acquired the *Star Mill* (*Frontier*) on the Yellowhawk which had been built in 1862.

The McCallay mill apparently was slated for oblivion by fire, for it again burned down in 1897, owned at that time by W.H. Gilbert. Mr. Gilbert did not rebuild the mill.

1. Walla Walla "*Weekly Statesman*," September 3, 1869.

Vanished Gristmills

WALLA WALLA COUNTY GRISTMILLS

MARCUS WHITMAN MISSION GRISTMILL

The gristmills at Waiilatpu at the Marcus Whitman mission, were not the same type of gristmills as were the ones built near the farm settlements for the farmers' use. The Mission mills were "tub mills." These were powered by horizontal water wheels immersed in a large tub-like container, and fed water falling from a short distance overhead. Whitman had three of these mills at his location and was planning a fourth when the massacre occurred.

Whitman's first gristmill, or tub mill, was in operation by 1838, and the grinding stones for this mill were wrought iron millstones, instead of granite ones. The iron millstones were only four or five inches in diameter and were "surrounded by a counter-burred surface of the same material."

This set of small, iron millstones was attached to a shaft of a horizontal water wheel, which took water from a chute a few feet above it. This put the water wheel in motion and turned the millstone.

Marcus got his idea for the gristmill from Spalding at Lapwai who had the same type of mill.

The other two mills which Whitman used at his mission in later years were larger ones, but still built on the same principle. The first mill was described by Thomas Farnham, (*Travels in Great Western Prairies*), who saw it in 1839, and said: "It was a crazy thing, but for it the doctor was grateful. It would, with the help of himself and an Indian, grind enough in a day to feed his family for a week and that was better than to beat it with a pestle and mortar."

At first, the American Mission Board in Boston did not think Whitman should be concerned with gristmills when he was sent out to Waiilatpu to convert and civilize the Indians, and they told him so. Marcus convinced them that before he could have any success with the Indians he first had to teach them farming methods and help them with their food supply, and having the means to provide them with flour was of great benefit, not only to the Indians but to the Mission as well. After that time, the board changed its mind and helped Whitman by searching for suitable mills to send to Oregon by sea.

Jason Lee, the head of the Methodist Mission in Oregon, was consulted by the mission board, and he looked into the matter of gristmills, supplying the following information:

Walla Walla County Gristmills

"I am now with Mr. Prichard (in South Lee, Mass.), the mill maker of whom I spake...His mills are not the same as the Ohio Patent, but act on nearly the same principle, and their weight and bulk less...Mr. Prichard has stones of the following dimensions, set in frames and may be propelled, either by horse, or water power. The largest 24 inches, second size 20 in., 18, and 16. Price of 24" with trundle head, ready for operation, one hundred dollars. The 16", price fifty dollars. Price of center discharge iron wheel and driving wheel to propell mill and boulting apparatus by water, thirty dollars."[2]

The board purchased a mill "designed for, but not including, 24-inch stones, and a boulting (sifting) apparatus without its cloth, designed for horsepower, to cost $125."

By March 1841, the second gristmill was installed and put in operation at Waiilatpu. William Gray of the Mission did the work of installing these mills. Later, he even made the millstones, chipping them out of native rock. He was more successful at this than Spalding at Lapwai, who found it "an extremely difficult task."

The stones for the second mill were of granite, and were each two feet in diameter, more than twice as large as the iron burrstones of the first mill. The mill was also run by water power, although it was also designed to be run by horse power if needed. They found this mill very satisfactory, and as Narcissa Whitman commented, "ample and safe, being altogether by a race (millrace) and not requiring any dam. The bolt (sifting frame) and all are in good operation. The mill will grind from one to one and a half bushels an hour. It is situated within a few rods of the house which makes it very convenient to attend to."

The mill served a good purpose and was much more satisfactory than the first "coffee mill" type. It was said they would have been able to supply flour and meal to both the Tshimakain and Lapwai stations if they needed any.

It is thought that neither of these mills were in a building, nor that the third was housed, being used out in the open. But years later, Gray, who was the "millwright" for the Mission, testified that the mill was housed in a wooden building about 20 X 35 feet. He also stated that in connection with the mill there was a threshing machine, so arranged that the wheat was taken from the field, placed in the threshing machine, threshed, cleaned and carried into a bin in the mill by elevators. The power for all this was supplied by the mill itself run by the water. It is thought that by the "building," Gray may have meant only a roof which was supported by posts.

2. Jason Lee, Aug. 9, to Green, (J. Orin Oliphant, *Lee-Greene Correspondence*, Nat'l Park Service Bulletin, 1973.

Vanished Gristmills

Sketch of third tub mill at Marcus Whitman Mission at Waiilatpu. Flume carries water from dam in left background to the waterwheel. Set of millstones at top of mill. Tailrace, or used water from the mill, in bottom foreground. Paul Kane artist.

Courtesy Marcus Whitman Mission, Waiilatpu.

The second mill was destroyed by fire in the fall of 1842, thought to have happened accidently, by an Indian boy. Everything in the mill was destroyed, including 200 bushels of wheat and corn and some flour. The bolting and the threshing equipment and even a part of the wheel were burned.

Marcus built his third mill in 1844, hiring a man to make a run of mill stones, the largest yet, forty inches across. He wrote to Boston requesting bolting cloth, a thresher mill and a corn thresher. Narcissa wrote that by October the mill was finished and that it "ground well." She stated that it was out of doors without a building. Statements of Marcus and Narcissa show their intention of erecting a building for the mill. Marcus even had a sawmill set up near the Blue Mountains in order to produce the lumber for the mill and other buildings he wanted to construct at the Mission. Lumber was being hauled to the Mission even up to the days before the attack by the Cayuses, with men busy at new construction. The attack and massacre took place November 29, 1847.

The gristmill was not destroyed by the Indians at the time of the massacre. Obviously, they wanted this part of the post to remain for their own benefit.

When Indian Agent Anson Dart visited Waiilatpu in 1851, several years after the tragedy, he reported that the gristmill was still in good condition but he was not impressed with the plant. He said, "The gristmill, as it is called, at Waiilatpu, is a very small affair, not at all entitled to the name, and would not have cost in the State three hundred dollars—I think it may have cost five hundred dollars."[3]

At the site of the Mission the outlines of the dam built for the third mill, and the mill pond can still be clearly seen.

The millstones have never been located for certain. It is thought that A.H. Reynolds when building the first mill in Walla Walla, the Pioneer Mill, secured "an old pair of granite burrs" from the mission site to use in the mill to grind "chop feed." No one has ever determined, however, if these were the mill stones from the third mill or from one of the others.

To this date, the Marcus Whitman Mission millstones have not been located for any certainty.[4]

3. *Anson Dart correspondence*, 10-20-51. Nat'l Archives, Nat'l Park Service, U.S. Dept. of Interior, 1973.
4. Whitman Mission Nat'l Historic Site, *Study and Report*, Thompson, Nat'l Park Service, Denver Service Center, May 1973.

Vanished Gristmills

WALLA WALLA COUNTY GRISTMILLS

ACME MILL at Lamar

The mill at Lamar, near Dayton, was built in 1898 by the Acme Milling Company. It was a large mill with a capacity of producing 200 barrels of flour a day. It was run by a turbine using water power, and equipped with Harding roller-process machinery. The Goss brothers were the first operators. It ground flour for only a few years, being discontinued in 1910 when it was acquired by Balfour, Guthrie and Company. A short time after the purchase, the building was dismantled and torn down.

* * *

According to the "Notebook of John White," from *The Times* of Waitsburg, small, privately owned mills also operated at different times along the Touchet Valley. In 1881, George W. Loundagin had on his farm on the Coppei Creek above Waitsburg a Corn Meal Mill with which he manufactured "*Columbia Corn Meal.*" He also had a castor oil factory on his place."

* * *

PHOENIX GRISTMILL

A *Phoenix Mill* in old Walla Walla County was built in 1872 and owned by J.W. Buchanan. It was run by water power with a capacity of 25 barrels a day. Said to have had two run of small stones.

Vanished Gristmills

Columbia County Gristmills

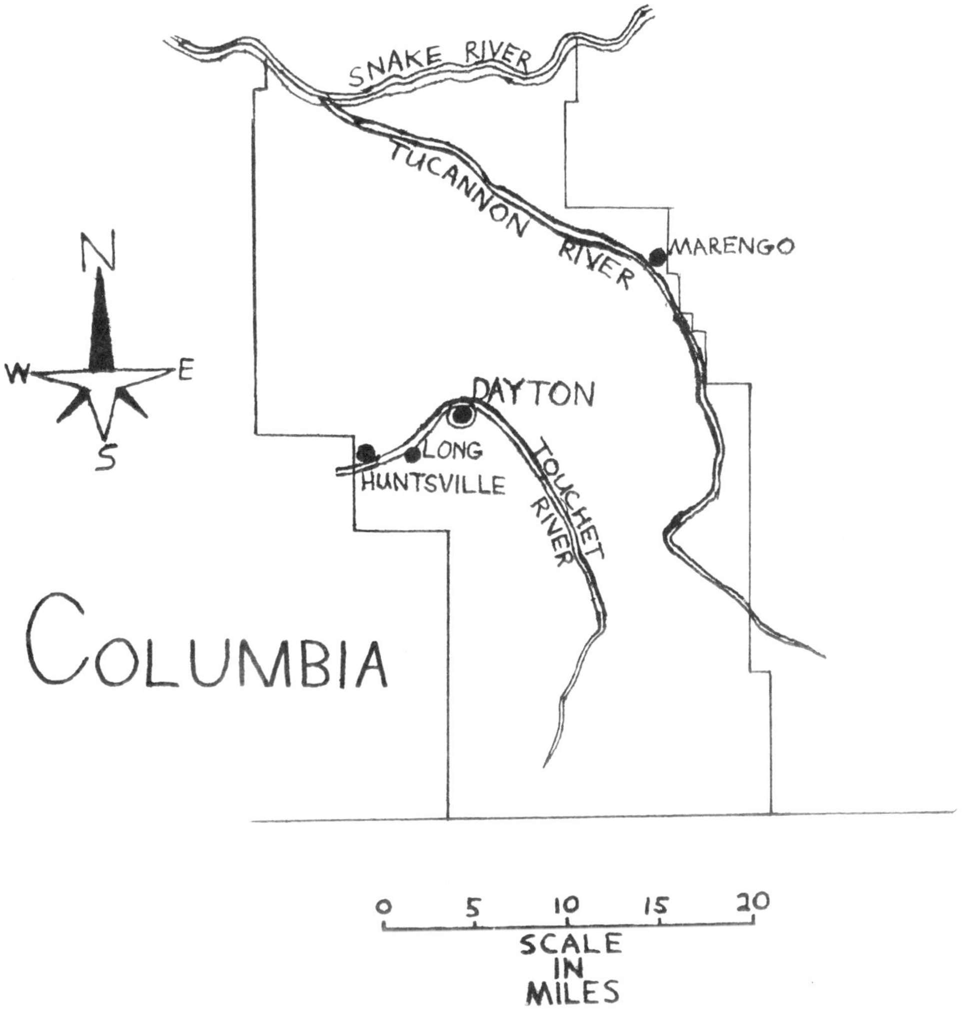

GRISTMILLS IN COLUMBIA COUNTY

Dayton (Wait's) Mill
Long's Mill
Marengo Gristmill
Corbett Mill, Huntsville
Brooklyn Mill, Dayton
Covello Gristmill

Vanished Gristmills

COLUMBIA COUNTY GRISTMILLS

DAYTON MILL

AFTER Sylvester M. Wait had sold his mill at Waitsburg in 1870, he started up the valley, perhaps looking for another location, having heard that Jesse Day was promoting a little settlement at his property on the Touchet River. Wait stopped and talked with Day. Day was an enthusiastic man who believed the little spot where he lived had great potential as a town. Right now about all it had going for it was a stage stop which Day was operating, and Inn to accommodate the travelers on their way from Walla Walla to Lewiston. But Day insisted nothing would stop a town from developing here if a gristmill were built.

Sylvester Wait knew Jesse Day was right, for there was no better place for an industry such as flour milling to take foothold than at a stage stop, especially if a strong stream was close by. He became an attentive listener and soon was as enthusiastic as Mr. Day.

Wait said if enough inducements could be secured in the way of land, water power, and rights as well as agreements from the farmers to support the mill, he would be willing to build one.

Day had these matters well in hand in just a few days. He donated five acres of land at the point "where the Kinney mill was later." He gave the right of way for the millrace and also the water rights, and included one block of land each to Wait and Wait's partner in this venture, Mr. Metzger, for residences. The farmers, eager to have a mill close by, gladly promised to bring their grain to the mill.

The gristmill immediately got under way, and at the same time Jesse Day platted the town of Dayton. As soon as the people knew about the gristmill and the new town, they were anxious to become a part of the enterprise, and began vying with each other for lots in the town to start up businesses. The gristmill began operating November 1, 1871.

Columbia County Gristmills

One of the Touchet Valley flou mills in the 1890s and 1900s Within a thirty-mile span ther were seven gristmills operating.

(Courtesy Whitman College, Penrose Memorial Library)

It is claimed that Wait took on the building of the $16,000 flour mill with $700, but that it cleared $5,000 in six weeks after it started running. According to one source, "In building the mill, with a brick building for a store and a planing mill, Wait and Metzger 'laid out' about $25,000, a large amount in those days."

The gristmill was water-powered by turbine with water from the Touchet River. Built before roller equipment was being used in western mills, this one was equipped with two and perhaps three run of granite stone, also bolting equipment and cleaning machinery. There undoubtedly were conveyor belts and chains for moving grain and flour to their respective stations, for Sylvester Wait would have built the best possible mill with excellent milling equipment.

At about this same time, the Dayton Woolen Mills was started, with A.H. Reynolds of Walla Walla, an early millwright and miller there, the chief owner. F.S. Frary, was the secretary and manager and Wait was president of the company. The woolen mills had a land site of seven acres donated by John Mustart and the building was erected at a cost of $40,000.

By 1875, Columbia County was formed and Dayton became the county seat which helped to spark its growth. The flour mill was a tremendous success, and according to Gilbert's *Historic Sketches*, "the mill was enlarged in 1882, equipped then with four-run of stone. At that time, a half interest in the mill was sold to S.H. Prather. It operated at a strong profit and changed hands a number of times. It was again updated and remodeled to the roller equipment in the late 1880s."

Vanished Gristmills

The mill was sold in 1883 to the Portland Flouring Mills, and later it was owned by Elmer Dunlap and Andy Berry. In 1954, eighty-four years after it was built, it was destroyed by fire at a loss of $50,000 and was not rebuilt.

The Dayton Woolen Mills was not a success and this industry closed down in 1880. The only woolen mills to operate successfully in this region was the Pendleton, Oregon, mill, as strong a business now as when first built.

Gristmills were so abundant in the Touchet valley at one time, they were only seven miles, or less, apart. But they each performed an important function with the farmers glad to have them located close to their farms and little towns, which were often only settlements. When the mills were built they no doubt were expected to be used indefinitely.

A farmer's world, and thus a gristmill's world then was not a wide one, but measured in the distances most easily and quickly traveled. An hour's ride by horse team did not cover much territory then, and it was a slow, tiresome hour, especially if the team pulled a large, heavy wagon load of wheat sacks to a mill.

Gristmills were continuing to be built even as railroad tracks were being laid into towns. Perhaps no one quite anticipated the dramatic change in transportation they would bring. Suddenly, there was a newer and quicker way to travel. And all at once the flour milling industry, undergoing a quiet little revolution all by itself, was producing newer and larger mills, equipped with the revolutionary roller process which made a different kind of flour from that ground out by millstones. Unquestionably, a fascinating, more exciting world was opening up which no one could have, nor should have, resisted.

There was no question that the world of the gristmills was suddenly passe, fading, and before long, vanishing. If anyone noticed, there was no concern, because the old could so easily be replaced with the new.

Columbia County Gristmills

LONG'S MILL

Barney Owsley, a well-known farmer and cattle rancher of pioneer days in the Walla Walla and Pomeroy country, said,[1] "...by 1866, enough farmers had located in the Touchet Valley to call for a gristmill. It was built at Long's, now the site of Long's Station a few miles below Dayton. The method of threshing then was primitive. I remember seeing Elisha Ping at his ranch on the Platt above Dayton, cleaning his grain in an old fanning mill after tramping it out with horses. I took grain to Long's mill in 1866 and had it ground and packed a whole train of it to Boise and another to Orofino (to the mining camps)."

This historic old *Long's Mill* was erected in 1866, but not entirely completed until 1869. It was a three-story wooden frame building, "about 40 X 50 feet, and 50 feet high, with big beams, hand-hewn tamarack logs." It was built on the Newt Curl homestead which had been acquired by Daniel Lock in 1866. The mill is said to have been built by Daniel Lock, John Long and Daniel Gilbreath.

There were two-run of granite millstones in the gristmill, and the mill could produce 100 barrels of flour a day, and if necessary, 125 barrels "in a pinch." The millstones came from Scotland, shipped 'around the horn,' first coming to San Francisco, then to Portland and up the Columbia to The Dalles. From The Dalles it was brought to the Long's mill near Dayton by freight teams.

It is said that the construction of the mill was spasmodic with farmers who were living nearby, doing much of the labor. They took this work out as "credit" with the mill for later milling services.

There was a simple bolting device for sifting the flour, but the bran and shorts were not used, their value as stock feed not recognized yet, and it is said these by-products were dumped into the creek.

The mill ditch carried the water from the Touchet River and the dam, about a half-mile to the millsite where it entered the wooden box flume. This flume brought the water to the overshot wheel. No evidence of the ditch remains for the land is now all under cultivation, but the wood from the flume was long ago structured into a hog shelter by the owner of the farm on which was located the old mill site.

The mill took the name of "Long's Mill" from John Long, and also his brother, Garrett Long. John raised cattle in California in the 1850s before coming to Eugene, Oregon. In 1863 he drove cattle into the Touchet Valley and apparently liking this

1. *Told by the Pioneers*, Vol. 2.

Vanished Gristmills

Two views of the early Long's Mill flour mill. Above is the early 100-barrel historic Long's Gristmill as originally built. This was just before it's change and enlargement. Note the two burrstones resting on the ramp. The lower view is after the mill was remodeled and enlarged. It was destroyed by fire in 1909. Courtesy Whitman College, Penrose Mem. Lib.

country, settled here in 1865 with his family, four miles below Dayton. Garrett settled in the Touchet Valley in 1866, and soon the two brothers, with Daniel Lock and D. Gilbreath, took on the building of the gristmill.

"At the time of the mill's construction, John owned 317 acres of land, two orchards, two residences, barns and other improvements. His brother Garrett, took up land near the millsite and made his home there."[2]

The mill through the years had several different owners at various periods, with Elisha Ping and John Long said to have been owners of it at one time. The mill was known as Milton Mills also, when it was owned by the A.J. Milton Company. The place where the mill stood has been known as "Milton," and some sources say Milton may have supervised the construction of the mill. In later years the O.R.&N. Company called it, "Long's Station," after they put in a side track at the millsite. The Gilbreath family is said to have been the main owners of the property on which the mill was built, throughout the years.

The gristmill was widely used in the region and farmers brought their grist there for grinding, for from the first it was known as a "first-class mill."[3]

But from 1880 to 1900 the old Long's Mill stood idle. The owners of the mill in this period were Willin and Church of Portland, Oregon, who seemed to have become involved in lawsuits, resulting in the property remaining idle during this time. As is always the case with man or machinery when unused and neglected, the mill and all the parts of its machinery and equipment began to decay and disintegrate. Anyone viewing the gristmill and the rusted gears and wheels would have considered it almost wreckage material, never able to be used in a mill again. Even the old millstones were decomposing and the mill race, neglected, filled up with debris and turned back to sod and grass.

In 1900, the Portland Company sold the mill property to Broughton and W.H. Gilbert. They went to work on the old gristmill, completely rennovating it and restoring it. They replaced the rotten parts of the structure with new timbers, and put the building back together again so well it looked not only as sound as ever, but even better, for now it was enlarged with additions made. A large receiving platform was built on the left side of the mill and a wheat warehouse on the right. The receiving platform had a huge grist bin into which the farmers' wheat could be emptied as it was hauled to the mill, and from the bin augered inside the building to be cleaned and readied for grinding into flour. The entire building was then painted which gave it a 'brand new' appearance.

2. Walla Walla, *Union Bulletin*. October 23, 1949.
3. Walla Walla *Statesman*, Sept. 3, 1869.

The interior of the mill was completely changed also, for the millstones were removed and a complete roller grinding equipment installed. All the cleaning apparatus and bolting machinery necessary for the work with the roller equipment was added. The mill, always an excellent one in the past could now again take its place with the best flour mills operating throughout the country, and there were many of them in this top class.

Shortly after it was rebuilt, Broughton and Gilbert sold the flour mill to A.P. Cahill who ran it for several years. But in the year 1909, the mill succumbed to fire, the fate of so many early flour mills. The handsome mill building which had been so carefully rebuilt was now completely destroyed except the warehouse. In later years this warehouse came under the ownership of the Touchet Valley Warehouse Company.

The land on which the mill once stood always remained under the ownership of the Lee Gilbreath family. This family is said to have broken up the one millstone saved from the old building, and pieces of it given to its members as souvenirs. The iron band which encircled one of the stones was once rescued just in time from a scrap iron drive to save it. It was given to the museum of the then Pullman State College.

Many farmers of the old days took grist to Long's Mill to be ground into flour. Barney Owsley was one of them, and his neighbors also hauled sacks of wheat to this mill, some carrying only one or two bags on the back of a horse, while others brought in good-sized wagon loads.

The Long's Mill was one of the very early mills built and for a long time seemed to have gone into oblivion, left for dead among a rubble of debris and ruins. But it was brought back to new life, and garbed in the fresh raiments of the roller process which all old-fashioned gristmills must surely secretly have coveted.

This new equipment brought extensive business to the Long's Mill again and extended its industrial life for several useful years. It would have continued grinding wheat for many more if fire had not sought it out and finally extinguished its life completely in 1909.

Columbia County Gristmills

MARENGO GRISTMILL

The little town of Marengo came into existence like so many others in that part of the country during the county-seat-scramble days. This was at the time Columbia County was formed in 1875 and a county seat for that county was needed. Dayton was the main aspirant, and Marengo became its competitor.

At first Marengo was just a little settlement on the Tucannon River, enjoying itself as an unspoiled spot of nature. It was deeply seated at the bottom of a steep canyon, surrounded by huge, luxuriant firs and pines, loved by the Indians who camped there in their tepees and favored by the many whites who had chosen it for their homes. It was on a direct path to Dayton and Walla Walla and people traveling there passed through it — but not willingly.

Physically, this terrain was a thoroughly hazardous one approached from any direction. Any farmer who must undertake its bumpy, narrow dirt roads with a heavy, unmaneuverable wagon-load of wheat on his way to a Walla Walla market, dreaded this part of his journey more than any other.

Still, when the new county seat election came up for Columbia County, this little settlement jumped right in, got itself made into a town — Marengo, and entered the race.

A number of good people helped get it started. John M. Silcott offered twenty acres he owned for a townsite to anyone who would start up businesses. This was the John Silcott who came into the county in 1860 to erect government buildings at Lapwai and who later married Jane, Chief Timothy's admirable daughter.

Mr. King, who operated the stage from Walla Walla to Lewiston, made a similar offer of lots at his place. But Silcott's offer was the one accepted, and as soon as the site was settled on, the name came next. Marengo was chosen to honor the Frenchman Louis Moragne' who had settled there about 1840.

Sewell Truax, the capable engineer of those times, laid out the little town. He is remembered for helping with the famed Walla Walla Baker Railroad, and for his first Pomeroy Wheat tramway down a steep bluff to the Snake River.

One of the very first major activities to get under way was the promotion of a gristmill. The residents went about this in a professional manner. They contacted a strong group known as *The Grangers*. This was a capable organization of seven grange groups in the county, the *Columbia County Council*, which was a joint stock company, managed by three representatives from each group. The organization recognized the need for a gristmill here but wanted to await the outcome of the election before going ahead.

Vanished Gristmills

When Marengo tried for the county seat post, its topographical location was against it from the beginning and Dayton won the election.

The Grangers, however, went ahead with their plans for the gristmill and built it. They began its construction in the spring of 1877 finishing it shortly thereafter. It was a good-sized mill with two-run of millstones and a capacity of 50 barrels of flour a day, and erected at a cost of $16,000.

As was always the gratifying situation when a settlement claimed a good flour mill as well as the added attraction of a stage station, the little town of Marengo grew rapidly and by 1878 a post office had been established, also a store, a hotel and a livery stable besides a school and a church.

By 1879, the flour mill had become a very successful industry in the town and countryside with many farmers using it and bringing in their wheat for grinding. That year it was incorporated and called, the *Patron's Flouring Mill Company*. Mr. R.P. Steen was the president, A.L. Sanford, the secretary and Thomas Throssell the treasurer. The mill was operated entirely by water power with a turbine using the water from the Tucannon River, then having "an unlimited supply."

The town of Marengo faded away before 1900 and there is little or nothing there to remind one of its former existence. Fields of wheat have replaced building sites. There is an old cemetery long forgotten, now covered with growing wheat, but some remember its location and can point it out to an interested stranger. There is an old stone building on a farm that once was a schoolhouse. Now it is used for farm storage but it has an unmistakable ancient look about it that harks back to the 1870s when it was filled with laughing boys and girls.

They, and the children of the Indian families who lived in the colorful tepees in the valley, played against the steep canyon walls. They all were fleet-footed as deer, loving their home set so deeply among the firs and pines.

Columbia County Gristmills

CORBETT MILL at Huntsville

At a site three miles up the Touchet River from Waitsburg, near the Walla Walla County line, a gristmill was built. This site became the little town of Huntsville, and the gristmill was built first to induce people to settle here and also start businesses and establish a town.

George T. Pollard donated the land for the mill and J.H. Richardson, a farmer built it with the aid of B.H. Hunt. Mr. Richardson, who was from Indiana, came here in 1863 and owned 760 acres of land in that region.

The gristmill was built in December 1881 and January 1882, and after it was finished, Mr. Richardson operated the mill until the fall of 1892. It was then purchased in the early 1900s by two brothers, A.E. Corbett and Judson Corbett. They owned and operated it from then until around World War I, at which time they sold the mill.

After the Corbett brothers died, the mill had many owners and at one time sold for taxes.

"John White's Notebooks" (*The Times*, Waitsburg), on early milling in the Touchet Valley is quoted as follows:

"The mill is only used at times to mill (stock) feed. It was purchased in May 1938 by Orville Fullerton and Sons, from Columbia County. Orville Fullerton sold his feed mill and the water rights that goes with the mill to the Ukiah Lumber Company in 1955, who is going to build a sawmill on the site. Fullerton kept the building across the railroad track for his apple packing house.

"On February 1958, the old Corbett Brothers Flour Mill was razed by the Dayton Lumber Company, (formerly Ukiah Lumber Company)."

Two other early items from the Waitsburg *Times*, indicate there were problems connected with the building of the Huntsville Flouring Mill.

"December 24, 1881—The flouring mill is almost completed, a few of the mechanics having left this week."

"January 21, 1882—The Mill Company have found numerous changes necessary to be made before the mill will be ready for operation. It does not pay to begin an undertaking without first qualifying one's self for the work, or employing someone thoroughly competent to take charge."

An advertisement in the Waitsburg *Times*, dated September 23, 1887, under Huntsville Flouring Mills, states: *"Richardson and Hunt, Proprietors."*

A similar advertisement one month later read, "*J.H. Richardson, Proprietor.*"

Another *Times* item, March 25, 1882: "The Huntsville Mill Company shipped its FIRST FLOUR this week."

An early Spokesman-Review paper stated: "Corbett Brothers, proprietors of the Touchet Flour Mills and Warehouse, have a gasoline grain stacker, the only one in this part of the county at Huntsville."

Built in 1881 and 1882, the Corbett Mill was powered by a turbine using the water from the Touchet River, and equipped with rollers. It was one of the strong mills of the Touchet Valley, much used by the farmers in that area.

* * *

Covello Mill

It is said there was a gristmill at Covello near Dayton. Sanford and Associates built the mill to grind wheat and cereals. It may have been mainly a chop mill although possibly ground whole wheat flours. It operated for several years. No other information regarding the mill is available.

Columbia County Gristmills

BROOKLYN MILL

Although Brooklyn is now a quiet residential district in the town of Dayton, at one time it was a busy little community with houses clustered near and around a gristmill. The gristmill at Brooklyn, built in 1882, was one of the much noted "Seven Flouring Mills in the Touchet Valley."

The seven mills were there, and a few privately owned other mills as well, such as a castor oil mill. The flour mills were Wait's Mill, Dayton (Wait's), Long's Mill, Prescott Mill, Huntsville, Brooklyn Mill, and the Acme Mill at Lamar.[1]

But perhaps having seven mills in a thirty-mile span in the Touchet Valley could have seemed an excellent arrangement to the farmers. In the 1880s and even in the 1890s, teams and wagons were still used and five miles took a good hour or more to travel. An hour in those days was no longer than it is now, and may have been, in relativity, shorter, considering that everything then must be done the slow way, by hand—or by foot.

Distances were thought of in the length of time and one's endurance for getting from one place to another. "Down the pike aways." "A hoot and a holler away." "A fur piece," or "As the crow flies," are colloquial phrases, now used in comedy for the laughs they bring, but at one time or another in many an old-timer's life they were a part of his language expressions.

Having a gristmill in one's community in the Touchet Valley less than four miles from another, or even in the Palouse country where there were many gristmills, could have provided the people living there with a certain measure of independence, offered a minor degree of status, and undoubtedly given a farmer and his family a comfortable feeling of security.

The Brooklyn Mill, (quoting from John White, in *The Times* of Waitsburg) "was built by two brothers, J.B. and W.A. Morris, along with two other men, W.S. Kinney and A.W. Sargent, operating under the firm name of Kinney, Morris and Company. The mill was 40 X 60 feet, with a basement, two stories and an attic. It stood down the Patit creek from Wait's Mill and near the mouth of the Patit. It commenced running with five wheat (millstones) and one barley stone, in May 1882. This mill ran for a short time but it failed financially, and lay idle for a number of years. It was acquired by the Portland Flouring Mills about 1900 and was destroyed by fire August 8, 1904, at a loss of $50,000."

1. *Union Bulletin*, Walla Walla, Feb. 6, 1938.

Vanished Gristmills

Whitman County Gristmills

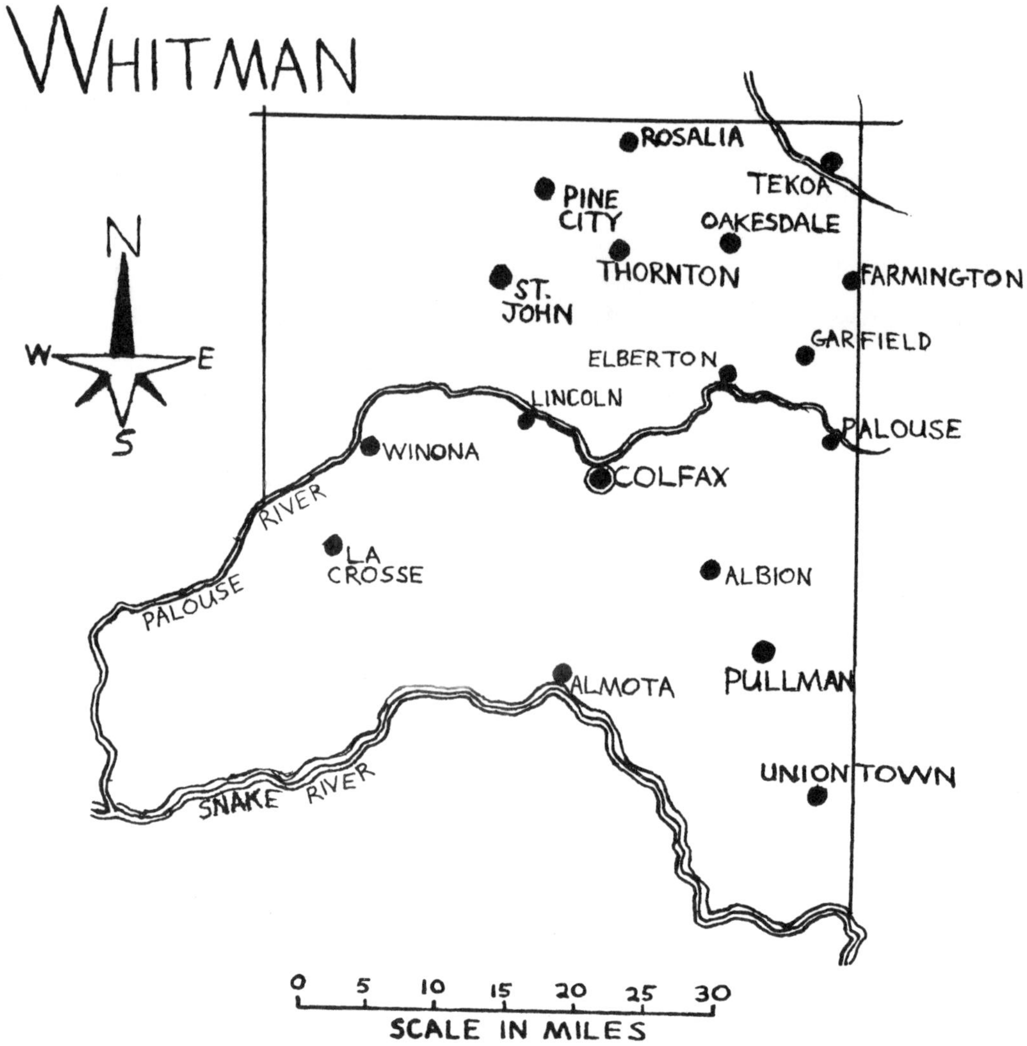

GRISTMILLS IN WHITMAN COUNTY

Palouse Flour mills
Colfax Flour Mills
Pine City Milling Company
Almota Flour Mill
Chase's Mill at Lincoln
Elberton Flour Mill
Oakesdale (J.C. Barron Flour Mill)
Garfield Flour Mill
St. John Flour Mill
Albion Mill
Winona Flour Mill
Farmington Flour Mill
Union Town Mill
Tekoa Flour Mill
Pullman Flour Mill
Rosalia Mill
LaCrosse Flour Mill
Thornton Gristmill

Colton Gristmill

Whitman County Gristmills

PALOUSE FLOUR MILLS

PALOUSE town was always well advantaged by nature because of the strong Palouse River coursing through its site. These rushing waters brought many benefits to the town and the people living there. Sawmills and gristmills, vital necessities in a new western town, were the first industries to take foothold along the river, and from these came improvements and developments which gave growth to the small town.

W.O. Breeding came to this area in 1874. He was a man interested in flour milling. He took into account the increasing numbers of homesteading farmers producing their excellent wheat crops, and measured the possibility of their need for a gristmill as well as the convenience of a nearby town as their trading center, located near the gristmill. He particularly favored a spot along the south bank of the Palouse. This was a flat area with easy access to the water, and an excellent location in his estimation for a gristmill. Before long, he decided to build one here and a short distance away, toward the west, a little town.

He immediately began the gristmill and secured a surveyor to plat the town of Palouse. By the spring of 1875 at least two businesses were under way in the new town. The operators of them were W. L. Powers, William Ragsdale, Waldnip and a Mr. Kelley. By the late summer of 1875, the gristmill was completed.

This new gristmill created tremendous interest, drawing groups of people from all around the countryside in admiration, to view all its complicated workings. A celebration in the form of a community gathering and dance was held in the big, spacious room of the mill, attended by young and old.

A dam held water from the river and this water was flumed to the mill to power it, by way of an overshot water wheel. At first, the mill operated with one run of granite millstones, the stones coming from France where most of the stones were obtained in those early days. The mill is said to have had a capacity of 30 bushels of flour a day.

Vanished Gristmills

Early Palouse Flour Mill in 1890s. Picture shows the dam as a Palouse family stands on a rocky ledge and views the scene.

Picture of the Palouse Flour Mill taken around 1910. Dam is shown at left of structure, also bridge.

Courtesy Paul Byers

Around 1880, a short time after the mill was built, it burned down. The blaze did not spread to other buildings in the town since the gristmill was far enough away to keep this from happening. The mill was rebuilt soon after the fire, using the pattern of the first mill and taking undamaged machinery and the stones from the original mill.

Another run of granite stone was now added and the capacity of the mill increased to 80 barrels of flour a day. To produce this quantity of flour, a turbine was undoubtedly put in to replace the water wheel. By the 1880s and 1890s, water wheels were not commonly used to power the mills as turbines were found to be better, giving increased power. The mill property now was under the ownership of J.H. Wiley, with the miller, F.M. Smith.

The Palouse Flour Mill had a simple bolting equipment for sifting the flour. A rectangular frame was covered with silk mesh, with a shaking device to jerk the frame from side to side as a moving brush underneath the silk removed the siftings. The siftings were the bran and shorts which were collected separately in different bins. This by-product which made good stock feed, was given to the farmers, included in the cost of grinding their flour.

As time went on, more and more mills found the demand for stock feed growing as fast as for flour. Many added chopping devices to their equipment just to prepare feeds for stock. This later became a big business in itself, often replacing the flour milling, especially in the smaller mills, when they no longer found it profitable to grind flour.

The gristmill had a steady stream of customers from the beginning. Some of the very first homesteaders who might have used the gristmill could have been C.H. Farnsworth, C. Oderlin, Lyle Hopkins, and J.M. Kincaid. Soon others, Henry Naylor, John Flint, Daniel Preffer, Levi Archibald, J.C. Cody, George Lamphere, Frank Normoyle, Morris, Curlin, A. Kincaid, W.H. and Joseph Lichty, Boyd and Cross, Cummings, Goodyear, Payne, Cornack and Wilson.

After the post office was established in Palouse in 1876, even more people were drawn to the region and the town itself began to boom. Some of the first merchants in the town were John Milhern, the Wheeler brothers who ran the hardware store, Fred Ankcorn, J.H. Follansly, J. Powers, S.J. Thornton, C.F. Robards, H.N. Boone, C.M. Meckles.

Jesse Bishop, an old-time miller of several mills in the Palouse country, was operator of the Palouse Flour Mills for several years.

"In 1881, Mr. Breeding died. "When the estate was probated, the land North of the Palouse River was dedicated to the city of Palouse, Washington Territory, as

Vanished Gristmills

Mill Dam. To the left, log race with logs falling over race. To the right, a fish ladder.
Courtesy Whitman Co. Hist. Soc.

NE side of town showing log flume with logs floating over race. Also mill dam, and fish ladder. Part of mill is on right and "old town" is in background.
Courtesy Whitman Co. Hist. Soc.

'Breeding Addition' to Palouse. The land continued from the North bank of the Palouse including the dam site and extension to the section line, but not including site of Palouse gristmill in inventory. The bill of sale of land was signed by the court and valued at $6,000. Later, in May 1882, this tract of land was sold to M. Cockerline and their son who in turn sold it to James G. Wiley, with the right to raise the existing dam two feet higher."

The above quotings are from the records of the abstract which Art Hunsperger retrieved from the safe of the mill after its last burning in 1923. Art's father was N.B. Hunsperger who was a pioneer miller in the Palouse country. He was with the Elberton mills for many years and came to Palouse as miller of the Palouse Flour Mills in 1905, remaining there as its miller and manager.

The fire which had taken this first gristmill, seemed to have been waiting, licking its chops for a bigger, better feast. In 1888, it swooped down and took its fill. It not only devoured the gristmill, but bent its ravenous appetite toward the little town of Palouse leveling it in one fiery gulp. This was a tragic occurrance in the life of the town and its citizens ever afterwards always referred to it as *The Big Fire*. Palouse has never forgotten the havoc it brought.

The town, like so many similarily afflicted at one time or another in their history, rebuilt. The rebuilding of Palouse town however, this time took place in another direction. People living there favored the north side of the river instead of the original south end. Eventually, this change proved a wise one and all the business houses were relocated there. This was done mainly because of the overflowing and flooding of the river which caused considerable damage at times. Now, the older part of Palouse is occupied by residences, and also the region's cemetery.

"J.G. Wiley and his wife, Ida, then bought the mill together, and later sold it to F.M. Smith, April 1889, who then sold it to Palouse City Roller Mill Company for $14,000, October 12, 1891. Palouse Roller Mill Corporation then sold it to J.C. Schumaker, April 26, 1898. J.C. Schumaker and wife Carrie sold to Wm. Voltz and Thomas Metcalf, doing business under the firm name of R.L. Voltz and Metcalf, on June 24, 1903."

The *Palouse Flour Mills* which had first burned around 1880 and again in 1888 was rebuilt, running successfully for several productive years as a roller mill. With this rebuilding, the mill was equipped with the roller machinery for making flour. A man by the name of Charley Towner is said to have installed the equipment. It is also said that J.C. Schumaker operated the mill as its miller for several years after that.

Vanished Gristmills

Picture of Palouse City taken in 1882. Then situated on hill south of Palouse River. After fire, many of buildings moved across river. (1) is General Merchandise Store owned by Wm. L. Powers and John Powers. (2) Hotel owned by Daniel Preffer, (3) Saloon, (4) Hotel owned by O.E. Clough. When they were moving it to new location, it fell through the ice. (5) C.J. Cox Home, (6) The "Fort," called this because it was once surrounded by a stockade. (7) Sam Dimmick's home. (8) Flour mill owned by Wm. P. Breeding who used water from Palouse River in foreground to turn overshot water wheel. **Courtesy Whitman County Hist. Society**

Lever's "History of Whitman County," gives a description of Palouse town in 1900:

"Palouse had a serious fire in 1888 and a set-back from the wet year of 1893. In 1901 the business places in town included, 3 general merchandise stores, 2 hardware and furniture stores, 2 drug stores, 2 second-hand stores, a candy factory and bakery, 2 butcher shops, one hotel in operation and one temporarily, 2 restaurants, 4 blacksmith shops, 2 wagon shops, a flour mill which is run by water power night and day when water is abundant; its capacity 80 barrels per twenty-four hours. A sawmill with a capacity of more than 40,000 feet per day, a sash and door factory, four saloons and a brewery, a bottling works, a bank, a cigar and stationery store, 2 shoe shops, a tailor shop, a millinery store, a suite of undertaking parlors, a harness and saddlery store, 2 photograph galleries, a newspaper, 5 doctors, 3 lawyers, and 2 dentists."

Whitman County Gristmills

In 1902, Mr. Schumaker sold the Palouse Flour Mills to Voltz and Metcalf of Kalispell, Montana, and they operated the flour mill with excellent success until 1905, at which time they decided to enlarge their business.

"Action was settled out of court by Washington, Idaho and Montana Railroad and Voltz and Metcalf were stripped of land along the North banks of Palouse River for track right of way July 25, 1905."

View of the mill from south side of the Palouse River. 1889. **Courtesy Whitman Co. Hist. Society**

The old mill was torn down in the fall of 1905, and a new three-story and basement building was constructed to replace it. The Barnard Manufacturing Company in Spokane installed all new roller machinery and equipment to go with it; a bran sifter, insulation and spouting, elevators, wheat cleaning and plan sifters. It was at this time that N.B. Hunsperger became the miller and manager of the newly equipped flour mill, coming from the Elberton Mill where he had been for several years. His son, Art, who "practically grew up in the mill" according to Palouse-

Vanished Gristmills

Town of Palouse, Washington. Population 2500 at time picture was taken in the 1900s. *Electric Flour Mill* on right. Courtesy Washington Water Power

View of Palouse country from top of Steptoe Butte. Courtesy Washington Water Power.

Whitman County Gristmills

View of south side ot mill from across the river on the north side. 1900.
Courtesy Whitman County Hist. Society

View of rebuilt roller mill from north side of Palouse River. 1920
Courtesy Whitman Co. Hist. Soc.

Pioneer, J.B. West, commented, "The spouting was wooden and the vertical elevators started in the basement, with a separate spout carrying the grist from each set of rollers to the top pillar where it was diverted to the plan sifter. Before grain was sent to the rollers it went to a vertical scourer to remove dust and smut, and ran to a horizontal conveyor and moisterizer before going to the rollers.

"The wheat raised in the Palouse area made a good pastry flour. It was Bart and Bluestem, and other wheat was shipped by rail from the Big Bend country and blended in with the Palouse Durum wheat in order to bring up the gluten content of flour for bread making."

Art said, "The mill was eventually bought by the Potlatch Lumber Company. They wanted possession of the log chute and tail race at the north end of the dam for their sawmill and planer, which was about a mile downstream from the gristmill. Most of the flour milled was sent to the Potlatch logging camps. Around 1911, a group of Palouse area farmers and a few Palouse business men, formed a company and purchased the mill from the Potlatch Lumber Company. Andrew L. Hill was elected president, John E. Turnbull, Secretary; N.B. Hunsperger, manager and miller. During World War I, numerous carloads of flour were milled and shipped to the Orient, mostly to China. All the flour was bagged in 98-pound sacks, the bags all furnished by the companies purchasing the flour. English only was used in the labeling on the bags, giving the name of the milling company that produced the flour. This was before the age of the electric sewing machines and all the bags for flour were of cloth, with burlap for feed and seed, and sewed by hand with the needle and twine. All hauling was by horsedrawn drays. The Palouse Flour Mill was not on a railroad, which made transportation more complicated."

On the night of January 22 and early morning of the 23rd, 1923, the Palouse Flour Mills caught fire and burned completely down. It was not rebuilt.

There was another mill built in the late 1890s, or early 1900s in Palouse, called the *Palouse Electric Flour Mill*, and was constructed by Risley and Wiley. It was about the same size and capacity as the Palouse Flour Mills and operated by William Gray and his two sons. It was said to have been dismantled in the 1920s and the site and buildings sold to the Palouse Grain Growers Incorporated who converted it to a bulk storage elevator for grain and peas.

Art Hunsperger recalled, "Not long after the end of World War I, the flour mills in the large cities took over and the small mills soon became a thing of the past. I can remember when most every town in the Palouse grain growing country had a gristmill to grind the farmers' wheat into flour."

Whitman County Gristmills

Wagon loads of grain hauled to Palouse Flour Mills for storage and seed cleaning. 1915.
(Courtesy Art Hunsperger)

Palouse Flour Mill flume and penstock carrying water to the mill's turbine.
(Courtesy Art Hunsperger)

Vanished Gristmills

All that was left of the Palouse Flour Mill after the fire of 1923.
(Courtesy Art Hunsperger)

The Palouse Flour Mills is an excellent example of the early water-powered gristmills. It's development kept stride with that of the town whose growth it sparked. Adversity in the way of fire, an unwelcome guest that visited too often, never succeeded in ending its work. The gristmill always managed to lift up its head and start over again, stronger and better than before. By adopting the roller process for producing flour it was able to continue grinding flour through the early 1900s, until fire again finally claimed it, for all time.

Whitman County Gristmills

COLFAX

A gristmill's beginnings more or less paralleled a town's growth and development, each in a way measuring the progress of the other. Such is true of Whitman County gristmills and the towns in which they grew and flourished.

As early as 1869, the Palouse country with its green hills and fertile valleys captured the interest of the new settlers coming West in oxen-driven wagons. Some of the first men who stopped and looked over this land, anxious to homestead and locate here, were Joseph DeLong, George Pangburn, William Castlemen, Philip Brogan, Nicholas Whelan, and McNeilly, among others. They liked this country and knew they had come far enough.

After securing their piece of land they began the arduous tasks of clearing; cutting pine trees and pulling stumps to get the soil in shape for planting a few acres of wheat. And they'd write to the folks back home urging others to come, telling a brother, a cousin or a friend, "sell your place and come West as fast as you can. There's plenty of land, this is a wonderful country." Those folks came, the brother and the brother's friend, or a friend and the friend's neighbor, many of them also settling in the Colfax area.

In 1870, Thomas Smith and James Perkins took land right at the confluence of the north and south forks of a steadily rushing river which the Indians called the Palouse. A short time later, Smith moved from his land, deciding he'd rather take out a claim nearer his Union Flat friends and relatives. In 1871, Perkins found another man by the name of H.S. Hollingsworth who took over Smith's property.

By 1871, this region was becoming well populated. Farms were producing excellent crops of wheat and every fall there was not only the marketing of the crop but also the urgency to lay in the winter's supply of food, the first item in that category being flour. Loads of wheat were driven over rough wagon roads by horse team to the gristmill at Waitsburg to be ground into flour. It was a long way to travel and farmers wished for a gristmill closer to their area.

There was also another problem which made life difficult at times for the people in this location. This was the placement of the county seat. Now at Colville, Pinkney City as it was then called, the great distance there being an inconvenience and hardship for the people in the Palouse country.

But in 1872, this farm-settlement country became organized into a new county, and within this county, a newly formed town became the county seat.

Vanished Gristmills

Anderson Cox, interested in this Palouse region, was mainly responsible for these developments. He knew the needs of the people living here, the most urgent, the matter of the county seat. He also wanted a gristmill built for the use of the farmers. Since he held the office of Receiver for the U.S. Land Office at Walla Walla this enabled him to use his influence in promoting a bill to create the new county called Whitman County, and the new county seat, the name of it as yet undetermined.

Mr. Cox next began working at plans to get the gristmill in operation. This gentleman, along with another interested citizen, D.S. Bowman, in 1871 met with three other men to discuss these matters. The men were James Perkins, H.S. Hollingsworth, and Levy Reynolds and the meeting was held in the sawmill which these three men owned and ran, having built it in 1870.

They discussed many issues of a nature which would benefit the people around there and promote the growth of the area, and before they were finished they had reached two decisions. One, that the new town which would be the county seat, should have the name of *Colfax* in honor of Schuyler Colfax, then Vice President of the United States; and two, that there should be a gristmill built in Colfax. They formulated plans on the best way to get the gristmill under way, deciding they should secure the pledges and promises of the farmers to bring their wheat to the mill and patronize and support it. Later, James Nosler worked hard all during the month of July in 1872 getting these subscriptions and promises from the farmers to bring at least 5,000 bushels to the mill.

The town of Colfax was platted in 1872 and businesses were soon under way with stores and shops being built and stocked with goods. James Perkins, whose land was suitably located where the north fork and the south fork of the Palouse met, gave the site for the gristmill. Anderson Cox took on the responsibility of obtaining the equipment and embarked on a trip to San Francisco to order and secure the machinery. Hollingsworth and Reynolds, who understood the workings of a mill's machinery, agreed to handle the operations of the gristmill after it was finished.

Unfortunately, an unexpected and tragic incident brought the well laid plans for the gristmill to a sudden stop for the time being. This was the death of Anderson Cox. He made the hurried trip to San Francisco for the machinery of the gristmill, but he was never to complete the trip for on his return, near Dusty, he was stricken with a heart attack and died.

None of the other men seemed willing to assume the responsibility and go on with the gristmill's construction after Mr. Cox was no longer with them to manage the undertaking. In early 1873, however, Mr. J.C. Davenport stepped in and took over the mill's plans and arrangements, assuming the responsibility for building the gristmill.

Whitman County Gristmills

Remains of early gristmill after a fire.
(Courtesy Bill Walters)

Under Mr. Davenport's management the gristmill, built at what is now the northeast end of Colfax, was completed and ready for operation by the end of 1873. It was a large mill with one run of stone, and a capacity of 50 barrels of flour a day, and water-powered. James Taylor is credited with hauling the machinery from Walla Walla and assisting in installing it. The first millers to run the mill were James Cooper and D. Wolford, both noted as "first-class millers," running the mill on a twenty-four hour basis.

The Colfax Flour Mill suffered a bad disaster in the spring flood of 1879, which not only damaged the mill but caused a life to be lost. An employee of the mill in attempting to cross the river by means of a rope, lost his hold when an ice floe struck him and he fell into the water and was carried swiftly away.

Vanished Gristmills

In around 1881, Mr. Davenport sold out his interest in the Colfax Flour Mill, moving to Cheney to run the gristmill there which he had built in 1880. The owners of the Colfax gristmill in 1882 were Livingston and Kuhn.

In 1882, The Colfax Flour Mill was apparently partially burned. An early history notes: "On July 14, 1882, a fire starting early in the morning, broke out and nearly leveled the business district of Colfax, destroying fifty-eight buildings....Livingston and Kuhn sustained a loss of $35,000, with insurance of $12,000." The mill was rebuilt.

By 1886 the Colfax Flour Mill was under the ownership of F.M. Ellsworth and William Codd. Under these men, the mill was enlarged and new roller equipment installed. The mill now had a capacity of 125 barrels of flour a day. Along with the flour producing equipment, a chop mill was added to produce stock feed which became a productive part of the mill's output.

The Colfax Flour Mills experienced a bad fire in 1920 but was rebuilt. It operated for thirty-seven years until a big fire in 1957 leveled the mill for the last time.

Another flouring mill operated in Colfax, constructed in 1892, built by W.H. Mastin and called the *Colfax Milling Company*. The Spokesman-Review noted: "The Colfax Milling Company is operating a mill built in the year 1892, using roller equipment."

Another Spokesman-Review item, dated August 13, 1911, stated: "The Colfax Milling Company adds a 56 X 76 foot extension to its premises for storage of flour and feed. A branch line of the Oregon, Washington R.R. and Navigation built a line to the mill. The mill is entirely overhauled and fitted with new roller equipment. The capacity will be 250 barrels a day. A complete oat and barley mill will be added to be operated by electricity so that all water power may be used for increased capacity of the flour mill."

The Colfax Milling Company was under the ownership of H.H. Warner and C.H. Warner. They manufactured a flour which they called, *Royal Rose Flour*, a patent flour, shipped to a wide market.

Whitman County Gristmills

PINE CITY MILLING COMPANY

What were the farms and the country like around Pine City in the 1870s and the 1880s?

The topography of the region then was practically as now. There were undoubtedly more uncultivated, tree covered hilly areas, considered either not worth planting or left as they were because they were too difficult to plow.

Many horses were a necessity on the farms. They were carefully cared for and fed much of the wheat and grain crops a farmer raised. People raised their own chickens, and their milk cows, and there were always the big, dirty pigs rooting in dirt or lazily scratching backs against pieces of a broken fence.

Farm houses were filled with large families of children. All the girls grew up to be good, economical cooks and were taught that men were the bosses, or at least should believe they were. The boys were raised under the adage, "Make hay while the sun shines," and told that days were intended for working, with idleness never tolerated.

There was Smith's gristmill at Pine City where a sack or two of wheat could be taken in at any time with Mr. Bishop having it all ground into flour in a few minutes. There was the security in having the town close by, with a doctor available who would come out to the farm in his horse and buggy even in the middle of the night if someone were sick and needed him.

Dr. Z.T. Dodson was such a doctor. With James Richardson and one or two others, he had come to this region in 1872. Dr. Dodson, recently out of medical college, was a man of vision, and the strong little creek coursing through the area interested him. He visualized a town here, with the creek, Pine Creek, used to help develop the town, perhaps supporting a gristmill. There already was a sawmill on the creek nearby, run by William Lawrence, which meant lumber was available for building. And there were plenty of places on the creek for a gristmill.

In 1878, Dr. Dodson acquired a large piece of property along the flat which included the creek, and the well-traveled low place through it, which was for many, many years an Indian crossing and which came to be called *Texas Trains*. Dodson lost no time in getting the site platted for a town.

Other men also were coming into the region, mainly from the Oregon country, and when word got around that Dr. Dodson was anxious to get a little town started along the creek, some quickly bought lots from him and started up a few businesses, while others took up farms in outlying regions. Anderson Edwards was one of the men who came up from Oregon and stopped here on his way through, looking for a promising place to stop. Edwards had up to now followed the carpenter's trade, but had it in the back of his mind to start up a gristmill in some suitable location. He knew nothing about milling but thought he could handle it after he had built the gristmill. The flour milling business looked like a sure thing to him and he was anxious to get into it. When he saw Pine Creek, noting its strong current and its ample supply of water, he realized this was a perfect site for a flour mill. Without further hesitation, Edwards sought out Dr. Dodson and before long he had bought from Dodson a little track of land on the creek on which he planned to build a gristmill.

Anderson Edwards
(Courtesy Henry Smith)

Acquaintances and friends as well as family members often followed an early settler to a new region. An acquaintance of Edwards by the name of Andrew Jackson Smith had relatives who had settled in the St. John country. Andrew Smith had been interested in this region since 1865, when he had come up from Oregon to 'look over the country.' A few years later, with relatives well established here, he came up again, liking it even better than before, and decided to move up to the Palouse country as soon as possible. He returned to Oregon, (he originally was from Iowa) and disposed of his property, "selling out for $5000 cash, which he turned over to his wife telling her to 'take care of it.' " She took care of the valuable money in the best way she knew how, painstakingly sewing it inside her corset which she wore constantly, even during the night, until the journey was completed.

Whitman County Gristmills

After reaching the Palouse country, Andrew Smith heard that his old acquaintance, Anderson Edwards was at Pine City and was starting to build a gristmill. Andrew saddled up his horse and rode over to see Edwards. He learned of Edwards' plans for the mill as well as Edwards' need for money, and evidently feeling the mill business in Pine City would be a good investment, he lent Edwards $1500 with the promise it would be paid back the next fall. By the end of the following summer, 1881, the gristmill was finished and the town was fast growing into a little city, with farmers around the area anxious to begin bringing their grist to the mill for grinding into flour. But now before business could get under way, there was the matter of the note due Smith. Edwards, since borrowing the money from Smith had acquired other unpaid debts having borrowed from other sources and now was unable to meet any of his obligations. Smith feeling there was no point in extending time on the note, foreclosed, taking the mill in default of the note and giving enough money 'in lieu of the value of the mill' to allow Edwards 'to pay off his creditors.'

Andrew J. Smith, owner and manager of the Pine City gristmill from 1881 to 1907.
Courtesy Henry Smith

"Suddenly," Andrew Smith's son Henry said, relating the story to this writer, "Andrew Smith found himself in the milling business." Henry Smith, the youngest of Andrew Smith's eight sons and three daughters, lives at Pine City and although the mill was torn down three years before he was born, he has all the facts and information well at hand regarding the mill itself as well as his father's work with it.

Vanished Gristmills

Not long after Andrew Smith took over the Pine City gristmill in 1881, another gristmill was started three or four miles below Pine City on the creek, built by a man named Henderson, who also ran this mill. But the mill, for various economical reasons, folded up soon after its construction and burned down shortly after. So now, according to Henry Smith, "The Pine City mill was without competition, the only other mill being a number of miles away at Post Falls."

Andrew Smith went right to work getting the Pine City mill operating at full capacity and hired a miller by the name of Hudson. Somehow Smith and Hudson were not always in complete agreement and this miller did not remain in Smith's employ for long. There were other millers who ran the mill; one or two names recalled by Henry Smith were Verstegg, and one of Andrew Smith's older sons, James Alva Smith, (Henry's brother), who learned the milling business, and the veteran miller of those early days, Jesse Bishop. Mr. Bishop also ran the Palouse City Mill and the Chase's Mill up above Colfax for a time. Usually, one or two boys were always employed in the mill to do odd jobs.

The old building had two stories and was built of lumber which came from the old William Lawrence sawmill, located about three miles away on the creek. The large 10 X 10 pine beams were hand-hewn but the rest of the planks and siding were circular-sawed, with square nails holding them secure. There was one set of millstones, and these buhrstones, 36 inches in diameter, were imported from Belgium, "shipped around the horn, to Portland and from there by river steamer to Texas Ferry (called Riparia now), then hauled to Pine City by team and wagon." The mill was powered by a turbine, called a *Leffel Wheel* and it was contained in a wooden enclosure. It was fed water from a dam one-half mile up the stream from which the water flowed into a ditch and then along an elevated wooden box flume which came to the upper middle of the mill building. There a wooden penstock, or standpipe, sixteen or twenty feet long, carried the water straight down into the turbine. The turbine took a sixteen-inch entry pipe.

There was no bolting frame in this early mill but one was put in later. There were no conveyor belts to carry grain into the building, or flour away from the stones. The flour after its one operation of grinding, fell into a wooden trough beside the stones, from which it was scooped up by hand and either emptied into a bin for storage or put directly into the waiting farmer's white, seamless flour bag. The output of the mill was sixteen barrels a day, or thirty-two, 125 pound bags of flour in twenty-four hours.

All the wheels and cogs of the machinery and equipment were made of wood. The turning and meshing of these gears gave forth a subdued, rhythmic "cush-cush" sound, further mellowed by the sheep's tallow which was regularly slapped on to all

Pine City gristmill as it looked in 1890 about ten years after it was first built. The construction on the right with the trestle supports was built for a track on which cars filled with wheat or flour sacks could travel to and from inside of mill. Later, an elevator was put in the mill but the building remained the same. **Courtesy Henry Smith**

the cogs to keep them running smoothly and prevent wear and breakage. However, if a cog should break or wear out, another was whittled out of wood and inserted into the socket which had held the other one.

The power was generated in this mill in the same way as in all early mills. It began from the long vertical wooden shaft extending from the turbine. At the end of this shaft was a wooden cogged wheel, called the master gear. This meshed with a smaller cogged wheel which had a horizontal shaft ending with another cogged wheel. From here another vertical shaft ended with a large grooved wheel which was belted to a smaller grooved wheel in a half-cross (twisted) connection. A vertical shaft from this ended with a small, cogged wheel which locked into the eye of the runner-millstone on the floor above. This turned the runner stone, but not the bottom, nether stone. Neither did the stones touch each other, for this would have ruined them.

The stones were adjusted just a fraction away from each other so they would never touch, which would mar the surfaces. They thus would cut grain, not smashing it, and gradually grind the grain into flour, the grain flowing out along the furrows in the stones to their feathered outer edges.

The flour from this grinding was a cream-colored, healthful, slightly coarse whole-wheat product, containing all the ingredients of the wheat grain as nature grew it. The stones grinding the wheat in this mill, like all other millstones in all other gristmills, had to be sharpened, *dressed*, frequently. This Pine City mill had a hoop-shaped device which could be attached to two holes at the sides of the runner, top, stone, and lift it easily, placing it in a convenient position for easy dressing. The bottom, or *nether* stone, of course, could remain where it was.

The sharpening tool, or millpick, could either be a wedge-shaped iron, its one end tapering to a smaller sharp edge, or a double-edged tool grasped by the hand. One wrist was held against the other so the strokes of the pick could be carefully controlled. Often, the miller dressed his own stones, but usually he had someone do this for him; a man called a "millpeck" who went around to the different mills doing just this kind of work.

Pine City gristmill millstones, now resting on a concrete base in yard at Perkins House Museum in Colfax, Washington. Note small hole in side of top runner stone. There is a matching hole on the opposite side. The hooks of the hoop-crane were inserted in these holes to lift stone to convenient position for dressing, or sharpening. Stones came from Belgium.

The water fall is the overflow from the ditch when the turbine was shut off. Water had to go somewhere so there was an overflow.

This is the wheel house. A shaft led from turbine to what was known as a head house, where a set of beveled gears conducted power into mill to operate the millstones.

Turbine located near bottom of this wheel house. Turbine was housed in a wooden box. This picture taken about 1893.

The addition on the right contained 3 wheat storage bins. At first, mill ground farmer's grist while he waited for it.

The hoop-crane certainly was a necessity in the Pine City Mill, eliminating a back-bending task of lifting weighty stones, but there were few enough other conveniences to save a back in the mill, or in other mills built then, for that matter. All the wheat sacks which came up to the loading door of the mill, had to be lifted from the big farm wagons by hand, hoisted painfully to the shoulder and 'toted' in, some of the smaller, younger men staggering under the 125-pound or more, weights of the wheat sacks. Henry Smith remarked that his six-foot, 240-pound father, Andrew, was a "big strapping man, strong as an ox." He could flip those sacks up to his shoulder as easily as if they'd been a sack of horehound drops from the corner store, but he was an exception. It was hard work for most, and after a full day of bringing sacks into the mill and bending weary backs once more to pack them out again as flour, just as heavy, but unwieldy and harder to handle, some men were "all in" at night. More than one back cried for a good rubbing with *Perry Davis' Pain Killer* by a concerned wife before her husband would 'hit the hay.'

But Smith thought up an easier way to simplify the sack-lifting tasks and likewise speed the work operations as well. He constructed a bridgework on the outside of the building with a little track for cars to bring the wheat from the farmers' wagons into the mill by way of the second floor. This arrangement worked out so well that before long Smith had an elevator also installed to carry these sacks of wheat and flour, up and down the two floors.

By 1890, the mill was doing a brisk business throughout the country, as well as in the immediate region, the farmers' trade known as "Custom" and the market trade, the "Merchant."

For many years the *Pine City Mill* was one of the most important mills in the Palouse region with farmers from surrounding areas coming long distances with their wheat for grinding. One gentleman of those early times, (Wm.Condit) Wild Goose Bill, who lived at the site of the present town of Wilbur, named after him, would come to the mill with a pack train and get a large load of flour. He'd take this wheat all over the countryside and sell it practically door to door. At Wilbur, he had a trading post and sold some of the flour there, and he also operated a ferry across the Columbia River where the Grand Coulee dam is now built, so he had a wide market and probably made a good profit from the sale of the flour he bought from the mill in Pine City.

A story is told of one customer who brought his wheat to the Pine City Mill to have it ground. He was in a big hurry and became dismayed to the point of angry frustration because he had to wait his turn from prior customers who had arrived ahead of him. In a huff, he grumbled loudly, "All right, I'll just take my wheat to Post Falls," and he strode out the door, jumping up to his wagon seat and jerking the horses' heads toward the Post Falls road. He must have had a moment of concern when he glanced up and saw threatening black clouds scudding about overhead, but sometimes fury can blur reason. The elements were not on his side, for a short distance from his destination a strong downpour set in which thoroughly soaked all his wheat sacks. When he reached the mill at the falls and Frederick Post looked at his wet wheat, that gentleman threw up his hands and exclaimed, "Too vet—cannot grind vet wheat. Clog tings up. Nein! Nein!" No amount of pleading would convince the miller it was good policy to grind soggy wheat and there was nothing for the farmer to do except drag himself up to his wagon seat again, flap the reins against the weary backs of the horses, click his tongue at them and bend their noses back towards Pine City.

Many hours later, the tired, unhappy, but still defiant farmer wound the reins of his team around the hitching post at the Pine City Mill and walked inside. He went up to Smith and said he'd thought better of the deal and would like to have him grind his wheat. One glance at the soggy wheat sacks brought a grim shake of the head. "Sorry. Too Wet. Can't take that through the stones. Get it dried out first." The farmer was now about to have a stroke and he shook his fist at Smith and shouted, "I'll never bring my wheat to any of your mills in this country again. I'll grind it myself first." And that's just what he did. He bought some kind of a hand mill and for ever after that he, with probably every member of his family helping, ground up his own wheat on his own farm and turned out his own flour.

Andrew Smith operated the Pine City Mill until 1895, at which time he retired and turned it over to the direction of his son, James Alva, who ran it until his marriage, around 1899. After that time, the father, Andrew, closed down the mill, selling it and the property to the Milwaukee railroad. This mill was no longer in business after that for another mill was constructed by Andrew Smith to replace it. But the old mill stood many years, the railroad finally tearing it down around 1930.

The new mill was built in 1895 close to the location of the old mill and erected by Andrew Smith's son Ira L., who also with his other two brothers had learned the milling business. The father, Andrew, helped in the financing of this second mill which was built on his own property. The building was of unusual stone construction. It used the native, columnar rock prevalent in that area which was often quarried and used in the construction of homes and business houses in the region. The Andrew Smith home was built of this beautiful rock.

James Alva Smith, second son of Andrew, who ran the mill until the year 1899.

(Courtesy Henry Smith)

Vanished Gristmills

Pine City, *New Era Roller Mills,* second mill built by Ira and Elroy Smith, with the help of Andrew Smith. Built of quarried stone of that area, 1895.

Courtesy Henry Smith

Another view of the *New Era Roller Mills* of Pine City, built in 1895.

Courtesy Whitman County Hist. Society

Whitman County Gristmills

People were no longer satisfied with stone-ground flour having had the taste and use of the whiter, more refined flours produced by the new rolling mills coming into use then. They thought this flour superior because of its lighter and better baking qualities, preferring it. The new mill when built was equipped with this roller machinery, and it was a large mill capable of a much greater output than the former gristmill.

But even though the rollers were installed in the new mill, the set of old millstones were kept and also used, grinding flour for some of the farmers and people in the region who still preferred the 'good, old' whole-wheat flour ground in the old-fashioned way, believing it not only tasted better with all the parts of the wheat in it, but "if it was good enough for pa and ma, it was good enough for me."

Some of the early farmers around the area who would have used the Pine City gristmill were: The George Clark Family, James Butler, George Dyer, P. Carlton, George Jones, J. Burkhard, Alex Hubbard, B. Naught, John Stephens, Wm. Davis, the John Costellos, the Ben Pangborns, Brainards, Enoch Harthill who bought the first threshing machine in the country, the John Jacksons, John Sanders, A. Siegels, Dallas Blackmans, John Moore, Chas. Peterson, Tim Lane, H. Ketchum, Welch, S.O. Allen, Geo. Foley, Ben Funston, Shan Spurgeon, A. Crithfield, A.J. and E.A. Stone, Kile, Norris Bell, H.H. Leadby, Chas. Burbank, George Barber, F. Jeffries, John Crutcher, Ed Bradner, J.D. Thompson, E. Cook, H.G. and Perry Hastings, Chas Weston, Al White, Bob Middleton, J. Swan, W. Moreland, O.R. France, Robert Dyer, A.E. Davis, W.G. Woodward, H.P. Keech.

Many other people living all around the country used the mill, making a longer list than this but with some names not available: the John Ledbetters, Fred Ogden, Garrett Baldwin, Jason Beaman, Harvey Ketchum, James Gordon, John Evans, Dan Howe, F. Best, J. Wright, D.J. Hasardm,G, Babb, O. Lindgren, Al Stevenson, O. Melhus, Tim Lane, J. Thompkins, Alex Hubbard, Wm Baline, A. Fees, L. Rawls, S.P. Aylsworth, Grandpa Gully, John DeYoung, Wm. Druffel, E. Hunsaker, L. Purall, G.B. Seward, W. Beck, G.W. Dyer, the Roders,R.J. Patterson, H. Burrows, A.L. Martinson, F. Wood, J. Squires, Newt Jeffries, Adam Kile, Frank Nicherson, the Harmons and A. Siegel, Milt Cook and the Allens, who came in 1880 to the place later called the Seaman place.

The mill was called the *New Era Roller Mill*, and was operated until 1907 by the Smiths, then sold in 1912 to Pete A. Wagner, who ran it until his death in the year, 1916. This was the same Pete Wagner who was associated for a time with Ole Dybdall in the Chapman Lake gristmill in 1897, and with Frank Martin in the Cheney Flour Mill.

Vanished Gristmills

The Pine City, *New Era Roller Mill* next came under the ownership of P.M. Clark who made an unusual innovation in its reconstruction. This was in the year 1916, and the reason was for greater power. He removed the turbine from the mill or at least discontinued its use, and installed an overshot, wooden water wheel and connected its shaft to the machinery to run the mill. The water wheel was fed water from the same wooden box flume which had carried water to the penstock of the turbine, but now it was altered enough to fit its new use of carrying water to the top of the overshot wheel, and down into its thirsty wooden buckets. The old Pine Creek had come back into its own at last after a long, idle wait, and now rose to the occasion full force, to gather its waters together and turn the water wheel in the time-honored fashion of a creek's waters serving mankind. It kept faithfully to its appointed task until the year 1921.

Ralph Smith. Ralph and Elroy ran the mill as a chop mill in later years.
(Courtesy Henry Smith)

But the creek's usefulness was pretty well over, and water-powered gristmills in the 1900s were already relegated to the past, belonging to yesterday's life and yesterday's needs. A new impatient generation was apt to scoff at the slowturning granite millstones, and at the inadequate water power to run a mill. Some were only laughingly tolerant of the speed-scorning, dust-covered miller, or the old-fashioned appearance of an old mill's interior. Nearly everyone was asking for the new white, fluffy flour from roller mills, thinking it better than the old.

In 1923, a Mr. Hegler came to Pine City and he bought the old mill from Clark. He lost no time in discarding the water wheel, immediately converting the mill to steam. But milling in Pine City by now had already reached and passed its zenith, and was in a final decline from which neither steam nor electricity nor any other 20th century improvement could have breathed life into it. The little village had long ago dwindled to a smaller size and it no longer could support a mill when other, better mills in larger towns nearby could so easily be reached in a farmer's pick-up or car. In 1928, Mr. Hegler accepted his defeat. The Smith brothers, Ralph and Elroy bought the mill back from Hegler hoping they might salvage it some way.

Elroy F. Smith succeeded Alva Smith and operated mill until 1907. In later years, took it again and ran it until 1943.

(Courtesy Henry Smith)

The mill was not dead yet. Under the capable management of the Smith brothers, Ralph and Elroy, the old mill took on an entirely new personality. It could have been considered, perhaps, as taking a step downward in the mill's social pattern, if a mill can be thought of having a social status, for now instead of a flour mill, it became a chop-mill, mainly grinding out grains and feeds for the farmers' stock.

It still had its millstones which no one had bothered to remove, and now these granite buhrs turned again to produce whole wheat flours and cereals under the label of "Health Foods," which found a ready market.

This business worked out very well for the mill produced at a small profit, but later, in 1941, the mill was finally closed down for good. In 1950, the railroad company tore the building down, removing the final vestige of its part in the industrial days of flour milling in Pine City.

The country around Pine City still looks much the same as it did in the 1870s and 1880s so far as topography is concerned. There are now larger wheat fields with almost all places cultivated, only a few steep sidehills left with pine trees still clinging.

The Pine City mill was one of the outstanding mills in the early days, working for the farmers' needs in the Palouse country. The Smiths who ran it were fine, exceptional men, led by the first Smith, Andrew, who left his mark of excellence on that region.

Whitman County Gristmills

ALMOTA GRISTMILL

Almota was a good-sized little town on the Snake River in the 1880s and 1890s. It was a beautiful little village, close enough to the river for the sounds of its waters to blend with the robin's melody, and the chukar's scoldings.

The town no longer exists as it was, but the site still bears the name, Almota. The river is wider and higher because of the dams, its song is slower and less buoyant, and the robin's trill and the chukar's tirade echo in other fields.

Almota still proudly points to the time it was one of the earliest places in Whitman County to be settled. It remembers well that its location on the Snake River made it the principal shipping point for the entire country north of the Snake River. It was situated half-way between Texas Ferry (now Riparia), and Lewiston, 39 miles away. The stage coach line between Colfax and Pomeroy crossed here at one time, and also the telegraph line built in 1880 from Pomeroy to Colfax.

In the early 1870s, two men came here. They were L.M. Ringer and Henry Hart Spalding. Henry Spalding, son of Reverend and Mrs. Spalding who were missionaries at Fort Lapwai, was born and raised among rugged Snake River bluffs. He is said to have secured title to Almota land first from the Northern Pacific railroad, later adding other properties around it. He and Ringer made developments on their properties and Spalding became a resident, remaining during his lifetime. Ringer moved on to another location elsewhere. Thus, the credit for the establishment of Almota has gone mainly to Spalding, he being noted as "Father of Almota."

Colfax was established before Almota, but Almota in many ways had an importance which the inland town of Colfax could not share, and that was Almota's advantage on the river for docking boats to and from Portland. A constant stream of transportation was carried on here. Goods came either up the river in steamboats from Portland by way of the Dalles, or were hauled by team and wagon freight over the bumpy, dirt roads from Walla Walla. But Colfax had another advantage which outclassed Almota's location, and that was its easy accessibility to new settlers, making it more desirable than Almota, although Almota's farm lands on the higher elevations were plentiful.

Almota town itself was unfortunately rimmed in by high-reaching bluffs, giving it limited space with little extra land for expansion, and this area near the river was especially suited for fruit orchards. Almota was an important, natural harbor, but only a few acres of land were available for the fruit farms and for the people to build homes there, also. Eventually one or the other had to give way before both the farmer and his trees would find themselves squeezed against the bluffs, crying for more room.

After the town was established in 1877, it grew rapidly. There were two hotels put up by Spalding and Ringer. Most important, an excellent gristmill was built in 1877 by the Adams brothers, with the help of Carter and Ringer. With a ferry also established in 1877, an important link for traffic was completed to the southern areas of the country.

From the very first, the gristmill was a highly successful industry in Almota. People came from long distances to have their wheat ground into flour. The mill was powered with water from Big Almota Creek which was brought to the mill by way of a long ditch, a considerable distance up the canyon. This ditch hugged the sides of the bluff until it stopped above the mill and there was formed into a good-sized dam. From the dam, the water flowed along a flume and then down through a standpipe to the turbine which powered the gristmill's machinery.

There were two-run, or sets, of granite millstones in the mill. There is some question as to the source of the millstones. Some think they came from the granite quarry nine miles above Almota which belonged to M.C. Moore. One source states they were fashioned by a firm in the East and shipped to Almota "around the horn." Perhaps one set, the smaller run, came from the quarry. Often one run of the stones was smaller than the other run and used for grinding corn and cereals.

When the mill was dismantled in later years after World War II, the millstones were given to a collector from California.

The miller for the Almota mill was A.C. Spengler who came from Iowa, but was originally from Germany, where he had extensive training in this line of work.

To learn the milling trade in Germany, an apprentice went from mill to mill, gleaning knowledge in the intricacies of each mill's operations, emerging as a well-trained and experienced miller. Such was the apprentice background of the many millers who came from Germany to this country. These people more often than not were regarded with love and respect, for themselves as well as for the significant contribution they made to the early development of the flour milling industry, and to the towns in which they lived. Milling was a time-honored occupation and old-time millers were more often than not given special consideration just to induce them to come to a community for their skilled management of a gristmill.

Mr. Spengler operated the mill at Almota for a few years and was a very successful and well-liked miller. But around 1881 he and his family moved away, first going to Albion to run the gristmill there, and then to Lewiston to run a bakery.

Many a small boy would always remember that he could visit Mr. Spengler in the mill and watch him 'tending the stones,' and let his eyes follow the flour dust spilling out from the grinding buhrs. He would never be shuttled away but would receive a warm smile from Mr. Spengler who liked children and always had time to explain the running of the mill to satisfy a small boy's mechanical curiosity.

The Almota mill after Mr. Spengler's departure, was taken over in 1882 by J.A. and H.H. Hungate, who operated it from then on, continuing with its capacity of 75 barrels of flour daily from its two-run of granite millstones.

In later years fortune smiled less graciously on Almota and the gristmill. A bad cloudburst in 1894 damaged the gristmill badly and it no longer functioned. The town by this time was dwindling in population and could not support the gristmill any longer. Nothing remains of the gristmill now except its site, for it was torn down in the 1940s.

Towns like Almota greatly add to the growth of a county. Here the wheat grew abundantly on the wide, high areas, and the low-lying acres near the river produced lucious fruits of many kinds. The sweet, juicy peaches in Almota's orchards reached large, beautifully formed proportions. People from many miles around would come every year for them, always amazed and delighted in their delicious perfection.

It is regrettable that such towns as Almota could not have continued on indefinitely.

Vanished Gristmills

WHITMAN COUNTY GRISTMILLS

CHASE'S MILL at Lincoln

Several years after Colfax had become established in 1872 and the gristmill there was finished and successfully operating, another small settlement above Colfax developed into a town. This was at a point twelve miles away on the Palouse River, and on the property of Charles S. Chase, designated as "being where the road crossed the river going south from Garfield to Eden Valley on the Colfax-Palouse City road."

In 1877, Mr. Chase built a gristmill on his property and in the fall he established a post office there too, himself the postmaster. The business of the town was not so rushing that he could not take care of the mill and the post ofice, and he was able to manage both occupations very well.

Farmers around the region found this settlement with its mill and post office a welcome convenience, and they brought their grist to the mill regularly. The town grew, with the addition of a sawmill and a school and a store or two. In the beginning, the settlement was known as Chase's Mill, but the name soon was changed to Lincoln.

For a time, Charles Chase is said to have seriously considered putting up a woolen mill on the river, for there was still plenty of power after the gristmill and sawmill had taken their fill. He did not take this on, however, probably because he was becoming too occupied with the gristmill's business.

A market for their flour had developed from as far away as Spokane and when the first large order came in from there, the Chase brothers, happy at the new business, harnessed up the team, loaded up the wagon with bags of flour and hauled the load the fifty miles or more to Spokane. When they finally reached Spokane after the long, slow, bumpy ride over the rough, dirt road, they realized they had well earned the money they had so gladly looked forward to collecting. But they continued with this business and trips to Spokane, not hiring the hauling done but making the trips within the family. Their enthusiasm one morning in January dropped to a cold low, to match the thermometer's marking which registered zero. They stamped their feet on their way to the barn, and hunched their shoulders inside their coats against the freezing air. Inside they unwillingly harnessed up the bay and the black and hitched them to the wagon loaded with flour. Somehow they made the long ride in the freezing weather to Spokane but that was the last load they ever delivered, cancelling all future orders.

The mill had one run of granite stone and there was no bolting equipment to screen the flour for a finer grade and sift out bran and shorts. A dam from the river supplied the steady fall (head) of water to the millsite. From there it was flumed to the overshot wheel, pouring water over the top of the wheel. A turbine may later have been installed. Jesse Bishop, a pioneer miller who also ran the Palouse mill at one time, and the Pine City mill, operated the Chase's Mill at Lincoln.

In the later 1880s, the ownership of the Chase's Mill, though still in the family, was under A.J. Chase and James Chase, brothers of Charles Chase, the builder.

At one time, in 1886, the *Colfax Gazette* had noted that the Chase's Mill and the gristmill at Colfax were both temporarily out of commission at the same time, with the farmers having no place near to take their wheat for grinding. The Colfax mill's machinery was stopped until an important part which broke was replaced with a new one from the East. The Chase's Mill found its usually dependable source of power—the Palouse River, suddenly failing them by dwindling down into a mere trickle scarcely strong enough to lick the pebbles of the river's bed.

The newspaper said: "The flour is scarce in the city owing to the closing of the Lincoln and the Colfax Flouring Mills. The former for lack of water to run it, and the latter for repairs."

The Chase's Mill at Lincoln was moved to Elberton in 1888 where it operated successfully there as a flouring mill for many years.

Vanished Gristmills

WHITMAN COUNTY GRISTMILLS

ELBERTON GRISTMILL — Known as *Hinchcliff Gristmill*

By the time the town of Elberton was platted in 1886, it already was a small community center. A sawmill owned by McCoy and Smith was operating. Several people were living near by and farmers on a number of surrounding acreages were producing excellent crops of wheat.

A gristmill had already been built nine years before (1877) by Mr. Chase, and owned by Charles Hinchcliff. This was located some distance toward the eastern part of the town of Elberton. After the town became established, the gristmill was moved to the west end, and located on the North Palouse River in 1888. The *Colfax Commoner,* W.T., January 6, 1888, said: "The flouring mill at Elberton ground its first wheat this week."

Sylvester M. Wait laid out the town of Elberton in 1886, naming it after his deceased son. Mr. Wait was a widely known and well thought of man of those early years. He built the mill at Waitsburg, owning and operating it as one of the most successful flour mills in the country. He also built the first gristmill at Dayton, this industry helping start the growth of that town.

Before the Elberton property was acquired by Mr. Wait, it underwent a number of changes in ownership. In the early 1880s, a C.F. Wilbur homesteaded it from the government. He sold the property to Luther Kerns a short time later, and Kerns then sold it to two brothers by the name of Moore. The Moore brothers sold the property to Sylvester Wait.

As soon as Mr. Wait platted the town and after the gristmill was moved, Elberton grew rapidly. This growth was spurred by the railroad's extending its road up the Palouse River from Colfax through Elberton, Garfield and Farmington. Soon, besides the gristmill, two general stores were in operation, also a post office, blacksmith shop and wagon shop, two large wheat warehouses and a good-sized livery stable. A church and school were important additions to the town, and well attended.

The gristmill was a wooden structure, three stories and a headhouse, high. A dam was built at the river and water carried to the mill by way of a wooden flume. From the flume, the water entered the penstock and ran to the turbine which powered the mill. The mill had one run of granite millstone, the two stones in this run each measuring around three to four feet in diameter. The capacity of the mill was seventy-five barrels of flour a day.

Whitman County Gristmills

Old Elberton Flour Mill built about 1883 by Charles Chase. This was first constructed several miles up the river from Elberton and in 1886, moved to Elberton. (Note)—Rope power transmission from water power wheel house to creamery building lower right, which ran the creamery machinery. Water for power of the mill came from the Palouse River at Elberton.

(Courtesy Art Hunsperger, Supplemental information by J. Weber Leid.)

Vanished Gristmills

In 1890, Mr. Hinchcliff hired a well-trained miller by the name of N.B. Hunsperger to run the Elberton flour mill. Hunsperger remained with the mill until 1897, a few years later going to Palouse to run the Palouse Flour Mills as its miller.

Mr. Hunsperger's son, Art of Harrison, Idaho, and J. Weber Leid recount an unusual use made of the old mill, which gave it an added duty besides its role of grinding flour.

Mr. Hinchcliff had built a creamery which was located next to the gristmill, and this was powered by the gristmill's machinery. In their words, "There were two large iron wheels with a long, large rope alongside the mill. The wheels were geared to the gristmill's water wheel and the long rope turned a wheel on the creamery and furnished the power to run the creamery."

The flour at the Elberton Mill was bolted to remove the bran and shorts. A simple cleaning apparatus put the wheat in good condition before it went between the millstones to be ground into flour.

N.B. Hunsperger, who ran the Hinchcliff mill at Elberton for so many years, was a man well qualified in the milling trade. He received his apprenticeship in milling in a flour mill in Ontario, Canada. He was born in New Dundee, Ontario, Canada, June 27, 1867, of Pennsylvania Dutch parents who were of the Dunkard religion which is closely related to the Mennonite and Amish. When Hunsperger reached the age of twenty-one and had completed his apprenticeship he came to Spokane. This was in 1889. He looked for work in the new flouring mills in Spokane, and was employed as a night miller under Mr. Samuel Glasgow, a well-known Spokane man connected with flour milling. This employment was undoubtedly in the new Centennial Mills just built in 1889.

In 1911, the gristmill was again relocated, now moved into the town of Elberton. Improvements were made at that time in the mill to update it, although it was already equipped with roller machinery. The mill was still water-powered, using a turbine and turning out more than one hundred barrels of flour a day. In 1924, the mill underwent another change, being converted to electricity, no longer dependent on water power. The flour mill, however, only continued in operation four years longer, 1928, for it stopped running then for lack of enough business to keep it going. The building is believed to have been torn down eventually, for there is no mention of its ever having been destroyed by fire.

Elberton made an early industrial name for itself, for besides its wheat farms and its flour milling, it had developed a flourishing fruit market. Many fruit orchards grew in the region producing fine fruit crops. A large fruit drying business went into operation here which not only gave extra employment to the people in the community, but encouraged farmers to plant fruit trees, mainly prunes, which grew

Wheat shocked in the early days of harvesting. Fruit trees grow in a nearby field.
Courtesy Washington Water Power Co.)

especially well here, on otherwise unused areas of their farmlands, along with their acreages of wheat.

It is said that the public school at Elberton at that time was built on an ancient Indian cemetery. No doubt, many a school lesson became a study of the long ago tribe of Indians who had lived there and had chosen that spot for their final resting grounds.

The Elberton school district employed four teachers which would have meant their school was a four-room building. This was considered a large one in those days, with three grades assigned to a room. This also indicated the population in the Elberton area was good-sized.

Elberton's unique diversity of agriculture, the raising of fruits along with its growing of wheat and other grains, gave it prominence in many ways among the Palouse towns. Its gristmill was one of the outstanding in Whitman County and still remembered for its period of importance to the farmers.

Vanished Gristmills

WHITMAN COUNTY GRISTMILLS

OAKESDALE FLOUR MILL

A large, three and a half story flour mill, with steam plant section and warehouse as wings to the building, was erected in Oakesdale in 1890, by J.G. Porter. This was a roller mill operated by steam. The fuel to generate the steam was supplied by slab wood, the waste parts of lumber from the nearby sawmills, used to stoke the furnace of the steam boilers. Later, in 1898, a crib elevator was added to the east side of the building. The old mill building at Oakesdale is still standing.

It is said that some of the machinery and even parts of the building came from an old mill in Illinois, shipped out here in 1890 and installed in the mill. This is undoubtedly so, for some of the machinery is obviously very old, and also marvelously well constructed. It is heavy and sturdy, with its wooden bearings in as good condition now as when new. A pair of very small burrstones, or millstones, are in the mill. They are enclosed vertically in a metal casing on which is embossed the patent date of 1871, the stones originally used in the mill to make graham flour.

The story of the mill's beginning unfolds in a few newspaper headings. August 3, 1890, the Spokesman-Review stated: "Mr. J.G. Porter starts construction of a flouring mill at Oakesdale," and on November 25, 1890, "The machinery was installed." By December 25, 1890, the paper noted: "The Oakesdale Flour Mill is running at full capacity." A January 22, 1891, item of the same newspaper read, "Additions are made in the Oakesdale Mill." On December 11, 1895, an item stated: "Oakesdale Flour Mill closes. Unable to operate at a profit. Price of local flour cut to meet price of foreign shipments." On March 4, 1898, the Spokesman-Review again reported: "Oakesdale Flour Mill closes for lack of patronage."

Such is the brief historical sketch of the Oakesdale Flour Mills up to 1898, its operation spasmodic to 1900 and a little beyond. Although the mill produced an excellent grade of flour and had a good business, it was plagued by financial problems, and a changing ownership. 1897 saw a sheriff's sale of the mill and in 1898 a final closing down, to operate only at short periods of time after that.

But in 1907, the mill changed hands once more when it was purchased by Joseph C. Barron, an experienced miller and millwright from the East. He immediately took charge of the flour mill and before long, had "put it on its feet."

Joseph's father, Moses Barron of Pennsylvania, where Joseph was born in 1866, owned and operated a water-powered gristmill there. Joseph as a young man wanted

Oakesdale Flour Mill, built in 1890, with building unused but still standing.

to follow a different line than that of his father, and studied for the ministry. But after finishing theological school, he decided after all to go into the milling business and operated the mill in Pennsylvania, working with his father.

In a few years Joseph married and took over a mill of his own in Crawfordsville, Indiana, operating it for three years. The westward movement was well under way by then, and Joseph and his wife Martha joined a group to the West, coming first to Asotin.

It is interesting to note that Joseph Barron became associated with the Asotin Flour Mills in Asotin, going in as co-owner with Mr. Vollmer, a banker and business man in Lewiston. This was in 1903. He was in Asotin for three years when hearing about the mill at Oakesdale, went there to investigate the situation. He immediately purchased the mill, and he and his wife and family moved there.

Soon after Mr. Barron took over the Oakesdale mill, it was converted to electricity, and the addition to the building housing the steam generator, was converted to living quarters for the family. Joseph updated the machinery by putting in four new stands of rollers and changing the system to ball bearings. Through the years other changes were made in the mill, updating and improving it. Its capacity was eighty barrels a day. A barrel of flour was 196 pounds, and a sack of flour in those days, weighed 49 pounds.

Bagging stations (there were four, 2 for feed and 2 for flour) where flour was sacked. Brands of flour were "Sweet Home," a patent flour; "Shield," and "G & G Best."

Whitman County Gristmills

J.C. Barron, who purchased the Oakesdale Flour Mill in 1907. Mr. Barron was a miller and millwright trained in Pennsylvania. He also studied for the ministry. His son, Joseph Barron, continued with the operation of the mill after the elder Mr. Barron's death. The mill is under the ownership of the younger Mr. Barron and still stands in Oakesdale, although it is no longer running.

Standing beside the two-stage roller cabinets. Mr. J.C. Barron is in foreground.

(Photos courtesy Joseph Barron)

Vanished Gristmills

Early farmers living all around the Palouse region who might have used this Oakesdale mill were the F.H. Browns, J.W. Salmon, Arthur Walsh, Presley Elkens, Wm. Brundy, Thomas Secrest, the Hodges, Crawfords, Fred Mills, W.H. Jones, Strattons, James McCoy, Chas. Wood, Wm. Morrison, R.W. Duncan, Ed. Crawford, Ed. Reed, F.H. Brown, the Hughes, A.H. Crow, Mr. Matoon, the Millers, Caves and Whaley families, Fergusons, Frank Rider, Hennings, Balthis, and Cashup Davis. Also the Green and Martin families, the Carlyles were early farmers, and the E.H. Hanfords, well known pioneers. The five Doneen brothers were there along with the Youngs, Wm. Greer, Thomas Norton, Johnstons, W. Walker and the Sains.

After J.C. Barron took over the Oakesdale Flour Mill in 1907, he soon eliminated the steam boilers and put in electricity. This is the way the mill looked in the early 1900s.

(Courtesy Joseph Barron)

Around 1916, Mr. Barron "traded his mill" for a large farm acreage in Montana. Somehow, from the beginning this farming venture, to quote Mr. Barron's son, Joseph, "turned out to be a fiasco." But matching this hopeless failure, was the Mill's sudden downward turn under its new management. In 1919, Joseph Barron reacquired his flour mill and started all over again, putting the Oakesdale flour mill back again on its feet, as he had done so successfully in 1907.

The mill made flour for the merchant trade and more especially for the custom trade of the farmers, working at this steadily until 1939. At that time the mill took on a new operation. Flour milling was turning unprofitable, gradually becoming an

Whitman County Gristmills

A maze of finely fashioned wooden spoutings carrying flour from different machines.

Mortised, tenoned and pinned joints throughout building. Beams 10 X 12".

Large, six-foot power wheel

Wooden pin ends showing through thick beams in manner of old time building construction.

Complicated system of belted wheels and gearings

industry for mills in larger cities that were taking the business from smaller community flour mills. The Oakesdale mill was now converted to seed grain cleaning and storage an enterprise which proved especially profitable during the war years and afterwards.

In the 1940s the elder Joseph turned his mill over to the management of his son, the younger Joseph, who managed it after that. Joseph Barron, Senior, passed away in 1955.

Other folks in the countryside coming to the mill were the Dunlaps, Miles, W.T. Thomas, Dan Fish, Robert Kilpatrick, Byron Denney, Thomas Akers, David Talley, Ray Peringer, Nesbit Cole, M.R. Fish, Paul Bailor, Gene Bower, Thomas Murphy, Ora Bernard, Thomas Cokley, Robert Murphy, James Bailor, H.L. McClure, John Wilson, John Hardisty, Cyrill Reed, Dr. Gaines, Yost Littleton, W.H. Ragon, Dr. Frank Cogswell, Cyrus Lake, Charles Taylor, T.J. Simpson, Edward King, Thomas O'Connor, Robert Stairer, Geo. Kreutz, George Young, Halbert and Carl Ellner, James Mustard, H.J. Riggs, Lee Blakemore, George Winn, Fred Miller, Wm. Nottingham, Thomas Hill,Norton, John and Henry Kelley, John and Elihue Scott, Frank Robinson, Joseph MacEachern, E.J. Tramill, John Clement, Wm. Lamb, and Virgil McCroskey.

The Oakesdale Flour Mill operated until 1960 when it was completely closed down, no longer able to operate at a good profit. The mill building still stands, never having had a major fire to plague or destroy it. It is a large, weathered, but important looking structure. It stands in a little town which keeps its storefronts of an earlier day, which gives it an air of stability and solidarity.

Mainly, the area now is a mixture of older well-kept up homes, and new modern-type residences, all well cared for with attractive gardens and flower borders. The mill looms sentinel-like, a reminder of other days well-lived in this Palouse-country town.

To visit the mill's interior, under Joseph Barron's guidance, immediately dispels any thought that the mill is old and useless. Here stands well-kept equipment, everything clean and neat, evoking admiration, interest and unbounded respect for the complicated machinery as well as for the skills needed to make and run them. Here the old-time principles of cutting, grinding and pulverizing grain into meals and flours, have evolved into the roller machines, which perform these same functions faster and more uniformly. The grain is ground by corrugated steel rollers,two each enclosed in a cabinet, each roller facing having a different-sized grading. As if turned out by a skilled cabinet maker, a maze of beautifully crafted square-shaped, tube spoutings, of varnished light-colored hardwoods, carry flour, or wheat, from one location to another in different operations. Automatic processes brought the

Views of Oakesdale Flour Mill which is still standing, but has not operated as a flour mill since 1939. Original mill building was without the addition in foreground. Line in siding shows where it was added.

wheat to the cleaning machines, the dust collectors, the separators, the washers, the scourers, the tempering bins and then to the grinders, the sifters, the purifiers and the bleaching apparatus. When the flour had gone through all its different stages it was ready for the bagging stations, where it was augered down into a sack, and the sack stitched at the top with a sewing machine. What a far and different process from that of the first gristmills that were built around the regions.

Perhaps the first thing noticeable in visiting the Oakesdale mill, is its cleanliness. It is airy, fresh smelling, spotlessly clean, the grey painted floor without dust or trackings. The equipment is in such good working condition, that a pulled switch will start any of it running.

The basement houses the big, six-foot power wheel which dominates the room. Its spokes are of metal but the rest is wooden, and the laminated wooden rim's surface measures fifteen inches across and about four inches deep. A heavy, ten-inch leather belt fits on this power wheel rim and extends about ten feet away to turn a smaller pulley wheel. A myriad array of different-sized belted pulley wheels extend from the shaft of the power wheel, the leather belts which once fit so snugly on them now hanging loosely down from idle wheels.

Mr. Barron said a bad accident occurred in this room in the early 1900s. A worker's loose smock caught in the belt of the big power wheel and before the machinery could be stopped, the man was killed, thrown against the huge beams a few feet from the wheel. It is fortunate that so few lives were lost in ways such as this, for the older mills had most of its machinery exposed and some fast moving parts could catch a man off guard and injure him, or end his life if he were not careful.

Although now retired, Mr. Barron still loves to make flour. As a semi-hobby, he has put a small mill into a building in his yard, and grinds whole wheat flour, cereals, and cornmeal—with the germ of the kernel still in it. This gives the cornmeal a wonderful flavor. He has a ready market for his products. He uses the hard wheats from Montana for his flours and cereals, and watches the protein and gluten content in order to get a high-grade product. The little room for his grinding is as clean andwell-ordered as the kitchen of the most meticulous housekeeper, and it is obvious that no mouse or insect would be happy here.

The old mill building at Oakesdale is no longer functioning and stands idle. It is listed in the National Register of Historic Places, which may help to keep it intact for future generations to glimpse back into the past and learn how flour milling was done in by-gone years.

Whitman County Gristmills

COLTON GRISTMILL

Colton at one time was a little settlement on Union Flat Creek. It became a small town when J.B. Standley platted out the property in 1882.

One of the first business buildings erected was the gristmill. This, as well as the main part of the town was built on property of the Becker brothers, and the farmers around the region, got together and raised $2500 to help Mr. Standley build the mill.

As soon as the gristmill was built and operating, the little town of Colton developed and grew also. This was the pattern of most early towns in Eastern Washington, and nearly every town's buildings at first usually clustered around or near the gristmill, the most important business establishment in those times.

The Colton gristmill had two run of granite stone. These stones came from France in 1880 and were shipped around the horn, up the Snake River to Wawawai in 1882, and hauled by team and wagon over the rough, dirt roads to Colton. Two trips had to be made to bring these stones, and the strongest wagon possible used, for each stone weighed about 500 pounds, and there were four in all.

Some early settlers of Colton were Dr. McCollum, W.A. Shelton, R.H. Duff and John Boyles. They all, and many more, had a large part in getting the town and the gristmill under way. There were three blacksmith shops in the little town, two hardware stores, a watch making shop, a meat market and three saloons besides the gristmill and some grain elevators.

Through the years the gristmill changed hands. In the 1890s Casper Hamp and G. Baumgartner bought it. In the 1900s Mr. Hamp became the sole owner buying out Mr. Baumgartner, and believing it would cost little more to begin with a new mill rather than renovate the old one, he tore down the mill and erected another one. In the new one he installed the roller process equipment, discarding the granite stones.

But despite the fact that the Colton mill was a new roller mill and milling a fine grade of flour, business to the mill was falling off. The population of the little town had dwindled by the early 1900's, many people having moved elsewhere.

In the 1950's, the Colton Flour Mill turned to another industry, that of processing feed for stock, becoming a Chop Mill. This industry was by now an important one which many of the flour mills turned to after the milling of flour was no longer profitable.

In 1968, McGregor Fertilizer Company bought the mill, running it for a time. Eventually it went out of business, and the building was vacated and later torn down.

Vanished Gristmills

WINONA

Winona had an excellent flour mill in 1903, powered by water from the Palouse River. A long ditch carried the water from a reservoir on the creek several hundred yards to the mill. The Palouse makes a sharp bend on its way through the countryside, and here the town of Winona was located. The picture printed here shows how it appeared in those early times.

The flour mill, called the *Winona Flour Mill Company,* burned down in 1918, although it had closed long before then mainly because there was not sufficient water to run the mill.

Winona in the 1880s and 1890s.

(Picture courtesy Perkins House Museum, Colfax, Washington.)

Whitman County Gristmills

There were many flour mills in Whitman County built before the turn of the century. Some built around 1900 were not water-powered, but run by electricity. Information on some of the mills built around 1890 has been difficult to find. Most of these flour mills did not operate for long, as flour mills, often converting to chop mills for processing stock feed.

Albion: A gristmill was built on the south fork of the Palouse River which flows through its town. At one time Albion was called "Guy," but this name was changed in the late 1880s to *Albion*, the town being in existence since 1871. It is located "half way between Colfax and Pullman on the south fork of the Palouse River." The south fork unites with the north fork at Colfax to form the main Palouse River. The mill was built around 1890.

Farmington: Had a roller mill built in 1884 by McGifford and Hicks, and operated by steam. Mill had a capacity of around 50 barrels a day. Burned down in early 1900s.

Garfield: Said to have had a flour mill built in 1889, situated on a branch of the Palouse River.

LaCrosse: Said to have had a flour mill built in 1900s.

Pullman: Had a flouring mill built in 1890. This may be the one owned by McKenzie and Baird and located on north Grand Street "where it crosses Palouse Creek." A Spokesman-Review article, 3-29-90 notes: "Phil Brener will use artesian wells for power to drive flour mill."

Rosalia: Flour mill built around 1900.

St. John: Had a flour mill built around 1900, said to have burned down.

Tekoa: Flour mill built in 1903. This mill shipped 1,000 sacks of flour to the West Indies in 1899, according to Spokesman Review, 5-26-99.

Thornton: Had a flour mill built around 1890.

Union Town: Had a chop mill which may have been a flour mill for a time. The mill was water-powered. There were numerous springs here and two creeks, Union Flat Creek and Cow Creek. Mill built in 1890s.

Spokane County Gristmills

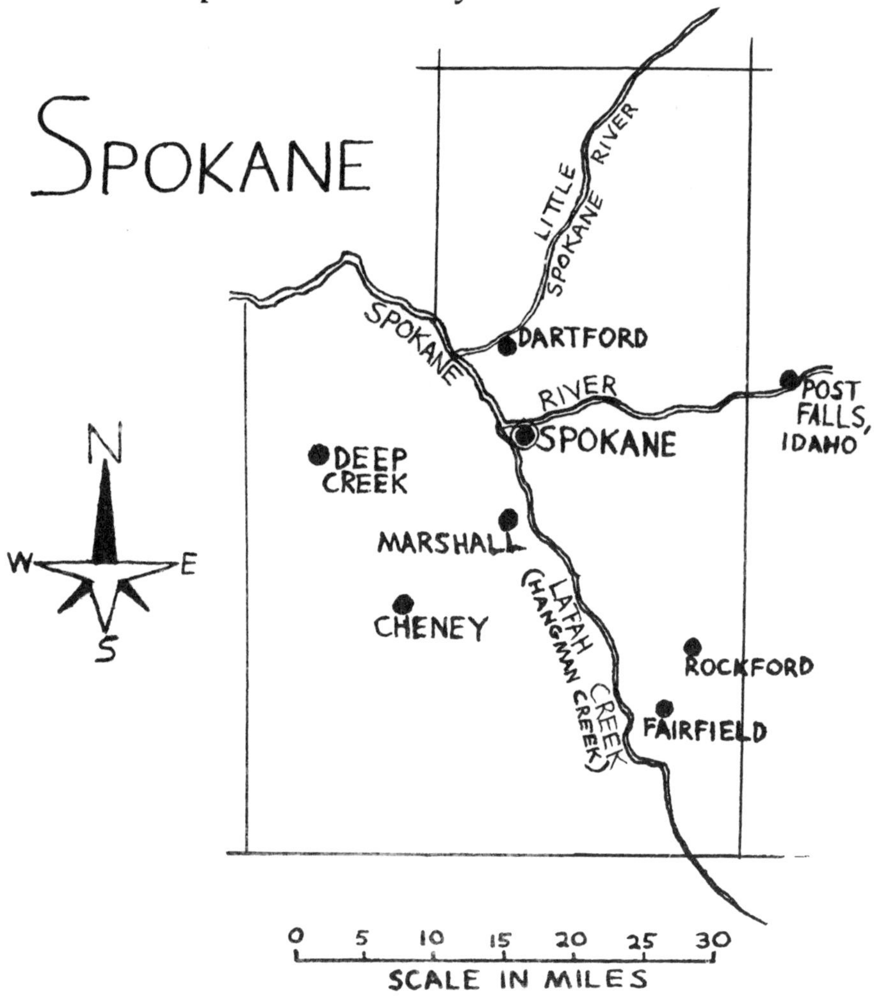

GRISTMILLS IN SPOKANE COUNTY

Posts's Gristmill, Spokane
C & C Flour Mill
Echo Flour Mills
Centennial Flour Mills
Dybdall Mill, Chapman Lake, Cheney
Dartford Gristmill, Little Spokane
Deep Creek (J.A. Bauman and Sons)
Marshall Milling Company
Joe Henry's Gristmill near Fairfield
Cheney Flour Mill
Cox's Mill
Rockford Flour Mill

Spokane County Gristmills

POST MILL in Spokane (Spokane Falls)

SPOKANE, OR AS it was first called, Spokane Falls, became established in the early 1870's, but was somewhat a late-comer in actual settlement. The plat to the city was filed at Colville in 1878, by James Glover. Glover, long called the "Father of Spokane", would have surveyed the town earlier, and made it into a townsite before then, but as he explained—"Money was hard to come by, and besides, there was no civil engineer at hand to do the surveying."

James Glover came to Spokane from Salem, Oregon, in 1873, with his friend N.W. Matheny. When he first saw the river and the falls he was completely entranced not only with their beauty, but with their great power potential as well. Almost immediately he was fortunate enough to come into possession of property along the river, which is now the main business center of Spokane.

Two men, J.J. Downing and S.R. Scranton, with R.M. Benjamin holding an interest, had "squatted" on an unregistered claim along the river, building a sawmill there. They had heard the Northern Pacific railroad would come through here and they hoped to cash in on the sale of lumber which would be needed then. The railroad did not come until 1881, and in the meantime the men became disgruntled with their mill which was beset with problems and mechanical difficulties, one of them being the saw was not adequate to handle big logs. They were more than anxious to get out.

As soon as Glover heard about the sawmill, he offered to buy out the owners. Downing sold him his interest immediately, but Scranton had to be hunted out, for at the moment had was in trouble with the authorities and was in hiding. He had obtained horses in a manner not in compliance with the law. Glover with help, secretly contacted Scranton who was more than eager to sell to Glover, so he could get out of town. After securing this property which joined the other, Glover was in possession of a valuable 160 acres, bordering the south side of the falls, and beyond. He quickly went into action to try to get a little town started.

He, along with Matheny and another partner from Salem, C.E. Yeaton, rebuilt the sawmill, increasing its capacity to more than needed then. He also built a house for himself , with a store at one end which he ran himself.

Indians were Glover's main customers at the store, and they found him an honest man when they brought in their furs to be traded for merchandise. He gave them anything they wanted except ammunition and whiskey.

One item on Glover's shelves was in short supply—flour—because it had to be hauled in from Waitsburg near Walla Wall, which was a long distance away. Glover was extremely desirous of having a gristmill built along the banks of the Spokane River which would use the power from its rapids.

The power from these falls was said to be 30,000 horsepower. Then, its use was limited to an industry which would locate on the banks of the river, using the falling water to turn a water wheel and furnish power for the plant. Although the first industries to use any early settlement's water power from a creek or a river was usually a sawmill, the more important and most needed industry everywhere was always that of flour milling, since it supplied a basic food.

Glover met and talked with more than one man capable of erecting and running a flour mill. John Houser who was with the Meyers Falls mill at Colville, and later at Pataha City in southeastern Washington, considered the possibility of coming to Spokane at one time. Glover offered him the site and water rights for $600. At that time, there were no wheat fields and the prospects for agriculture were dim in the Spokane region, while the fields near Walla Walla were already turning out good crops. Houser chose Pataha City instead for that reason.

The determining factor which kept many farmers and agriculturally-minded people from being interested in Spokane was its soil. It was gravelly with little clay subsoil. One man took a bucket of river water and poured it down into a little hole he dug, to test its water retention. The water disappeared as quickly as if it had been poured down a drain. The farmer threw his bucket to one side and exclaimed, "Nothing will grow in that kind of ground."

A Frederick Post, millwright and miller originally from Germany, had recently come from Illinois into the Rathdrum country in Idaho. Near him in Idaho, was a river, the mouth of the Spokane, and its crashing falls. He was as intrigued with this site as Glover had been with the Spokane Falls. But certain rights to the Idaho river and falls were claimed by the Coeur d'Alene Indians, which the government recognized then. This prevented any white man from building near the water, or using it, at least until that matter was resolved.

Post suddenly found himself as popular and sought after as an affluent maiden with an opulent dowry. Every young town throughout the entire region courted him,

Spokane County Gristmills

Frederick Post, who built and operated the first gristmill in Spokane. He also built and owned the first gristmill in Post Falls, and the first sawmill there.
(Courtesy Spokane Public Library)

offering him inducements galore if only he would come to that town and build a gristmill. Post not only had in his ownership most of the equipment necessary for an entire gristmill, but also possessed the ability to build and run one. This machinery he had brought with him to use whenever he was satisfied he had found the right place to erect the gristmill.

Tempting offers came constantly, but Post, a quiet, kindly-natured, home-loving man, would think about them, then talk it over with his wife, "Mama, what do you tink?"

"Its's so nice here, Papa. Let's vait a while, yeah?"

Post had five lovely daughters, and he wanted the best environment possible for them as they grew into womanhood.

Vanished Gristmills

Glover had met Post a short time after he came here from Illinois, and had given him several good offers to start a mill at Spokane Falls with Post usually countering, "Nein, nein. This iss a goot place, right here in Idaho. Maybe I build a mill here soon."

Glover knew if he ever got Post to come to Spokane Falls he'd have to present him with such a tempting offer, he couldn't possibly turn it down.

In the meantime, Post had started a gristmill at a spot— now Trentwood— "where the Northern Pacific crosses the river nine miles east." After working at the mill's construction, according to Glover, "all one summer, the next fall and winter and quite late in the following spring, during which time he was building a house and digging races for the mill, he concluded to give it up."

In Glover's own words from a March 26, 1917, *Spokane Daily Chronicle* account: "Post came down here to see what I could do for him, and I received him with open arms. I was all alone then and ready to make any sort of terms I could in order to get him to come and build a mill.

"I gave him 40 acres of my 160 after he had agreed to come here. I supposed he was more of a mill man than he proved to be later, and that his machinery was more up-to-date. I found it to be quite old-fashioned with the sort of (mill) stones a man would sit and peck at all day when he dressed them.

"As a matter of fact, the mill never amounted to much of anything. It ground out a little flour (its capacity was about 20 barrels a day), but very little wheat was raised in this part of the country then. The grain crops were mostly oats and barley.

"The 40 acres I gave to Mr. Post was west of Post street. The south line was Main avenue, and it extended across the river. I didn't include in it the block on which the Auditorium building now stands, because I had taken a fancy to it myself, and wanted to hold it. But Mr. Post wanted to buy it from me.

"I like that block so well because it would square my property out", he said. Finally, I told him he could have it for $350, and made him a deed for it, but didn't record it, as it would have been a little risky trusting it to the mails in those days and they were pretty careless with things up at Colville, the county seat.

"Not long after that, Mr. Post got strapped. He was building his mill and race to carry the water and needed some more money. I was getting in a little money now, although I was scraping every way I knew to get a dollar. Mr. Post came to me and said he was afraid he wouldn't be able to pay for the block, after all, and wanted me to take it back.

"A little while afterward, when he thought he saw his way clear again to pay for it, he came back and asked me if he could still have it. I gave him back the deed, and I suppose that block changed hands four or five times in that way before he finally paid for it. Later he sold it to Cannon and Brown for little or nothing.

Spokane County Gristmills

Post Mill **built on the bank of the Spokane River and end of Post Street. First gristmill built in Spokane, begun in 1876 and finished in 1877.** Eastern Washington Historical Society Museum.

Vanished Gristmills

Early view of Spokane Falls. Bridge across river is first Post Street bridge, and the two mills situated at the end of it are the *C & C Mill,* and the smaller one, the *Post Mill.* To the left is pictured the *Echo Mill.* (Courtesy Washington Water Power Company.)

"When he got his race built and his water over to the millsite, he asked me how he was going to get the gravel sluiced down to bedrock for his mill foundation. I told him I would do it for him, and it took me only three or four hours. I helped him with the construction of the mill, too, when he would need it.

"It was late in the fall of 1877 when he got it finished, and he had no wheat to grind and not much of a mill, after all.

"He and his wife and five girls lived in a house he built at Front and Post. It stood there for years. The mill was down below the house, in the block where the big power house of the Washington Water Power Company now stands.

(The house built by Post was the first two-story frame building in Spokane.)

"About 1880, Post sold his millsite to Jonathan G. Clark, (the father of F. Lewis Clark), who bought it for his son. Young Clark had as a partner a man named Curtis and the old *Post Mill* later was succeeded by the *C & C Mill.*"

By this time the town of Spokane Falls was developing fast, and Mr. Glover was busy with many activities. The Post Mill, apparently never a first-class gristmill in the first place, already was considered out-of-date only a few years after it was built. New ones were ready to take its place, for the power of the falls was unlimited with plenty of space available for other flour mills, now the leading industry at Spokane Falls.

Post moved back to Post Falls (the site there taking his name in later years). The Coeur d'Alene tribe during the years held steadfast in their claim of possession of the falls. Many misunderstandings arose with the white people as well as with the government, but Post always remained on the friendly side of the Indians. He went out of his way to be good to them, making them gifts and giving them provisions, or assisting them with other kindly acts, all done with the sincerity native to his personality. In later years when the government was negotiating a treaty by which the Coeur d'Alenes were to be placed upon the reservation, the Chief remembered the favors done them by Mr. Post and they insisted the agreement should hold the condition that the ownership of the falls belong to Mr. Post.

Mr. Post built a gristmill at Post Falls at the site he loved so well. He also built the first sawmill there. In much later years, when gristmills on the banks of streams and rivers were no longer profitable to operate, Mr. Post sold his interest in the falls to the Washington Water Power, who, as an early historian states, "harnessed the beautiful sheet of white foam to electric generators."

By 1885, Mr. Post had sold off all his Spokane property. The Posts were still fast friends of the James Glover family and were in Spokane often. Mr. Glover in the Chronicle article, reminisced of the time when Mr. Post was selling his Spokane property.

"Mr. Post and his wife came to my house to arrange for the sale of his last property here. He was away fixing that up in the evening and Mrs. Post was at our house.

"Mrs. Post," I said to her, "I am going to ask you how much Mr. Post has received from the sale of the property here I gave him."

She started laughing and said Mr. Post had said on his way to Spokane that he intended telling me that evening. He had it down in a little book she said, but she didn't remember how much it was. When he came and had been told of our conversation, he gave me the amount. It was $87,300."

(It is said that when he sold the power rights in the Post Falls, he received $22,500, reserving 100 horse power of electricity for lights and power.)

Mr. Glover continues, "We had never had a contract or anything but a verbal agreement. I had given him the property if he would come here and make Spokane his home and build a mill. I never blamed him for returning to Post Falls for he never would have been happy anywhere else."

Spokane's flour milling industry soon became one of major importance in the fast developing city. Frederick Post's gristmill started this, and also the development of the town, as James Glover knew it would.

It is good to know that Spokane's first industry, like Spokane itself, was built on trust in fellow man, when a hand shake was a binding contract, and when warm, honest business relationships as well as true personal friendships, were first in importance.

Vanished Gristmills

SPOKANE COUNTY GRISTMILLS

ECHO MILL in Spokane

From an early drawing. The mill on the Island in middle of picture is the Echo Mill. The mill at right is the C & C Mill at the site of the Post Gristmill. **Courtesy Spokane Public Library**

The *Echo Mill* was the second flouring mill built at Spokane Falls, and it was built by S.G. Havermale and George Davis in 1881. The first mill was built by Frederick Post.

Mr. Havermale and Mr. Davis were said to have had "limited funds" but with financing were able to put up the mill. It produced its first flour in 1883. Two years later in 1885, the basement of this mill was used as Spokane's first electric power house.

In 1886, the mill was enlarged and improved, and in 1887, Havermale and Davis sold the mill to B.B. Bravinder and A.E. Keats.

The Echo Mill survived the great fire of 1889, but in May, 1892, the mill caught fire and burned. The owners immediately rebuilt the mill, putting up a new structure of brick, at a cost of $50,000 which did not include the roller equipment and machinery put in the new mill. It was back in operation in a year which

Spokane County Gristmills

Upper Spokane Falls, showing Echo Flour Mill, California House, Glover's stable, Glover's sawmill on river bank, and flume of first electric plant in foreground.
Courtesy Eastern Washington Historical Society

indicates the importance of the flour industry in the city. By this time, there were two other mills, the C & C Mill and the Centennial Flour Mill, both producing flour on a wide scale.

Around 1900, complications arose, for the Hypotheek Bank which held a mortgage on the Echo Mill for $75,000, put on at the time it was rebuilt, foreclosed in the spring of 1899. At that time it closed down its operation.

According to records, "When the Great Northern was purchasing a right of way it bought the sheriff's certificate of sale of the property from the Hypotheek Bank. Shortly after that, Charles Sweeney, holder of the second mortgage, redeemed the property. Suit was then brought by Mr. Sweeney to restrain the road from grading the island or crossing the river at the millsite, alleging it would injure the water power for the mill. The Great Northern also brought suit to condemn the right of way over the property. The suit was dropped when the Great Northern (through its representative, Jim Hill) bought the property in 1900."

In 1901, the Great Northern leased the mill to the Portland Flouring Mills Company. Later, it was acquired by the Washington Water Power and under their ownership the mill produced flour until after World War One, around 1920.

The Echo Mill after that time remained idle for a few years, and in 1927 was demolished and torn down.

Early fire engine in Spokane

Spokane County Gristmills

C & C FLOUR MILLS in Spokane

The *C & C Flour Mill* was the third flour mill to be built in Spokane Falls and was constructed in 1885.

Although Frederick Post's gristmill, the first one in Spokane Falls built in 1879, was only a little less than ten years old by 1885, it was even then termed "outdated". The new process of using the rollers to produce flour was already in strong use in most flour mills, accepted as the best method of milling flour. The machinery for powering Spokane mills then was ordinarily turbines which used the water power provided by the cascading falls of the Spokane River.

The name "C & C" originated from the first letters of the names of the two men who built the mill—F. Lewis Clarke and Frank Curtis.

Clarke, born in Bangor, Maine, in 1861, came to Spokane after graduating from Harvard in 1884, choosing this part of the United States for its climate, believing it would be beneficial for a health condition plaguing him. Wanting to become settled in a business, and flour milling being the main industry of those time, he contacted Frederick Post, offering to buy him out. Post accepted the offer selling to Clarke, turning over to him the Post gristmill, the mill site and the water rights.

Samuel Glasgow, one of the founders of the Centennial Mill in Spokane, remarked in his manuscript memoirs, "I came to Spokane in 1882 as a young man to seek my fortune, and F. Lewis Clarke who already had a fortune, came West in 1884 to seek better health."

Clarke, with his partner, F.E. Curtis, a well trained miller who had been with the Pillsbury Milling Company in Minneapolis before joining Clarke, immediately began building a flour mill on the site beside the old Post Mill which they left standing. Clarke offered his friend Glasgow employment in the new flour mill, and Glasgow gladly accepted.

The C & C Flour Mill was a large one, 70 X 100 feet, five stories high "with fire proof machinery." It was said to have been the largest flour mill in the entire northwest at that time and put up a cost of $45,000, and including the machinery, $65,000. It contained the new "noiseless roller machinery," and was water powered by means of a turbine. It is believed to have been the first flour mill in the area to send flour to the eastern states and to the Orient.

Picture of town two days before Great Fire of August 4, 1889. The two large buildings are the C & C Flour Mill and the Post Mill beside it. The dark buildings at left are the Echo Flour Mill. All buildings to the right areas were destroyed by fire. Bridge leading from mill is Post Street bridge.
Courtesy Washington Water Power Company

The old Post mill beside the C & C mill, was turned into a cereal mill, both mills operating and producing high grade products, the C & C mill, flour, and the Post mill turning out oatmeal, pearl barley, hominy and buckwheat flour.

According to a very early Spokesman-Review article, the C & C Flour Mill was ready by August 1, 1885, just in time for the wheat from that year's harvest. It is stated that wheat "began to roll in from as far west as Waterville and Douglas County, 150 miles from Spokane." Glasgow remarked, "It took a week or more for some farmers to make a round trip when coming to Spokane to the C & C mill. They were delighted to receive forty cents per bushel for their wheat, going home with the winter's supply of groceries and clothing."

Frank Curtis, the other partner, with his extensive experience, set forth finding new markets for the C & C flour. He traveled everywhere, even abroad and the products of this mill went to many markets in the East as well as to Europe and the Orient.

The C & C Flour Mill operated for five years under the ownership of Curtis and Clarke and in early 1895, was sold to a strong local firm, called the Washington Water Power and Milling Company.

When the C & C Flour Mill was acquired by the Washington Water Power Company, it was said to have furnished that company's principal revenue at that time. Although the Washington Water Power Company was an electric utility even then, in 1891, it was also in the milling business. In 1893, the Washington Water

Picture taken in 1899 showing construction of a power plant by Washington Water Power Company below the old original C & C Flour Mill. To the right and beside the C & C stands the old Frederick Post Mill built in 1879, and small building at left of mill is Post residence —"first two-story residence in Spokane." Mills were destroyed by fire in 1904.

Courtesy Washington Water Power Company

Power increased the mill's capacity from 250 to 700 barrels daily. This was said to have been done for the purpose of selling the mill, which they did in 1895 to the Portland Flour Mill, a Portland based company which owned nine flour mills around the country at that time.

The Portland Flouring Mill Company, with George Shiel as the Spokane agent, ran the mill for nine years. The brands of flour manufactured during its ownership were the "Plansifter", "Superb", "Spokane," and the "C & C".

By 1900 wheat was brought from railroad box cars to the mill in the mill's electric cars. The cars carried the wheat into the mill and emptied it into bins without any handling by workers. By 1900, automation in the early industries and on the farms as well, was suddenly taking a firm foothold developing rapidly from then on.

Although all mill owners hoped one fate would never overtake them and their mill, the threat of this dread which was fire, was constantly guarded against. Yet it often came when least expected. On August 4, 1904, the fifteenth anniversary to the day of Spokane's Great Fire of 1889, the C & C Flour Mill and the old Post Flour Mill beside it, both caught fire and burned completely to the ground.

Vanished Gristmills

When the C & C Flour Mill burned, there were still two other strong flour mills operating in Spokane, the Echo Flour Mill and the Centennial Flour Mill. Now with the C & C mill gone, these two mills found their own business greatly increased and their mills working to full capacity day and night, to fill all orders.

At the time of the C & C fire, another fifth major flour mill had just begun its operation in Spokane. This was the *Spokane Flouring Mills.* The building of this mill company had been put up in the early 1890s, and stood on the north bank of the Spokane River opposite the site of the old C & C mill. It was built by Samuel Oppenheimer, who, as one historical source states, "got over-extended, and it was more than thirteen years after it was built, before the mill building was used to produce flour."

In 1903, the Spokane Flour Mills was leased to a Dutch firm represented in Spokane by Lodewyk Leus and it was then it began milling flour. This large building on the north bank of the river was seven stories high and made of red brick. The property included the Phoenix sawmill, the land on which stood the Centennial Mill and the water power from the adjacent upper falls.

The mill operated under the Dutch firm until the 1930s when it was purchased by Mark Miller of Portland who was operating a mill in that city. Mr. Miller felt the need to increase the Spokane Flour Mills' capacity and he brought many big machines up from his Portland mill installing them in the building.

Spokane Falls in 1880
(Courtesy Washington Water Power)

Under the new ownership the mill operated under the name, *"C & C, the Spokane Flour Mills"*, running all through the years at full capacity. After Mr. Miller's death in Portland, the mill continued to remain in the family, with Mr. Miller's daughter, Mrs. C.N. Smith president of the Spokane Flour Mills. Anthony Anderson and Edward Musgrove of Spokane were also connected with it. The mill during the years was said to have been enlarged, modernized and refurbished but happily never had a major fire. It stands intact today, an attractive and imposing structure in the cityscape, bordering the river.

The mill stopped milling flour the summer of 1972. That June the mill closed down for economic reasons, one of them "the exorbitant freight rates". It was at this time taken over by its Spokane creditors.

By 1972, the city of Spokane was becoming "World's Fair Minded", for this major attraction was to be held in Spokane in 1974, and all thoughts were turned to making this an outstanding, exciting and successful event. The old mill building was to play a colorful role in the many changes in preparation for the fair, and soon the building was under extensive remodeling, mainly its interior. Its seven stories were to be used as offices, shops and restaurants, the mill building still retaining the atmosphere of an old flour mill, turned into an attractive center of its own, drawing

From an early drawing. At right C & C Flour Mill with part of Post Mill at right next to it. In foreground of C & C Mill stands Frederick's residence. Large building at left is Echo Mill. Bridge in foreground is Post Street bridge.

An early picture of the second C & C Spokane Flour Mill, a seven-story brick building on north bank of Spokane River. Washington Water Power Company power plant at right.

(Courtesy Washington Water Power Co.)

crowds of shoppers and visitors to its countless attractive shops and eating places. The building kept its original design and structure. The old massive beams were left open and as they were,to be viewed by the admiring public. Parts of the basement wall were left exposed showing its thick, seven foot width of basalt and mortar, and the high ceilings still showed their stout wooden beams and bracings, and occasionally an iron pulley.

This was all finished in good time for the World's Fair held in Spokane in 1974 and proved a huge success. It has remained as a permanent shopping point in Spokane and is a popular, much-used enterprise.

Structures such as old flour mills, old courthouses, and even old residences are more and more given serious consideration for their historic and structural value and appeal before they are rudely knocked down by the wreckers, for a new building, or more often, a blacktop parking area. Some of these old structures are just as sturdy and strong now as they were when first built, or can be made so with some minor restorative help. They are all well worth preserving and so is their history, and nowhere is history more deeply imbedded than in the first flour mills, which were the basis of the flour milling industry.

Spokane County Gristmills

CENTENNIAL MILLS IN SPOKANE

In 1889, George Pahl of Avaca, Iowa, owner of the Centennial Mills in that city, became interested in starting a new flour mill in Spokane. He was encouraged in this by his friend Samuel Glasgow, who at the time was with the C & C Flour Mills in Spokane. The two men formed a company and built a flour mill in Spokane, the fourth one built in that city. They called it *Centennial Flour Mills.*

The flour mill was completed by August 3,1889, and on that day its new rollers began grinding wheat into flour, producing its first one hundred barrels that day.

On the very next day, August 4, 1889, the worst fire in Spokane's history raged through the young city—the Great Fire. The Centennial Mills was not destroyed in this fire, and neither were the other flour mills standing then. It is interesting to note that those first one hundred barrels of Centennial's flour was distributed to the needy after that grave event.

From the very first, business poured in strongly to the Centennial Mills, and within a year's time it was necessary for it to double its capacity. Again, in another few more years, the output was doubled. At this time, an oatmeal mill was built

Spokane's Centennial Mills, in 1889

beside the main flour mill, and cereals such as pearl barley, oatmeal, and other products were manufactured there.

An early Spokesman-Review article made this comment about the Centennial Mills: "The machinery is of the very finest quality, the rollers being of the latest improved pattern, and this mill boasts of being the only full roller mill in the northwest. Palouse country wheat only is used and the company finds a ready sale for its entire output, shipping as far east as Chicago. Operating its own elevators this company is enabled to select only the very finest grades of wheat. The officers of the company are: M. Thomsen, president, with O.F. Hall, vice president. All grades of feed, grain and mill stuff are manufactured and dealt in, and a general milling business done."

Centennial Mills during the early 1900s, branched out its operations and built mills in other cities under its name. There was one in each of the following cities: Tacoma, Seattle, Wenatchee, Ritzville, Pasco, Sprague, Reardan, Vancouver, Creston, and Wasco, Oregon.

The milling industry all through the years proved highly successful for the Centennial Mills. A pie factory in Chicago alone, took as high as two cars of flour a week.

On August 2, 1902, (August seeming to have been a favorite month for Spokane's fire monster), the Centennial Mill, and its cereal mill burned. The mill, however was immediately rebuilt, a five-story, brick building which took the place of the former old mill building. It operated for several years until November 7, 1909, at which time it again burned down, at a loss to the owners of $200,000. Both of these fires had started in the cereal mill, the second one with what all millers dread most—an explosion.

The mill was again rebuilt on Howard Street, and remained standing until torn down.

An entirely new Centennial Flour Mills was built in 1939 on East Trent. This location could be used because water power to run the mill was no longer needed. The headquarters for the Centennial Mills later was in Portland, Oregon, but it was not the same firm that was in Spokane in 1904.

The Centennial Mills built in other cities were finally closed down: Seattle was dismantled in 1917, Sprague closed in 1922, Reardan, 1927, Portland was converted to a feed mill in 1931, Ritzville closed in 1945, and the Tacoma mill discontinued because of fire in 1947.

In 1900, Spokane was said to rank seventh among all the mill centers of the nation.

C & C, the Spokane Flour Mills **building at left.** ***Centennial Mills*** **at right. Elevators at back of Centennial Mills advertising Big Loaf Flour, undoubtedly the name of a brand produced by Centennial Mills.**

Courtesy Robert Weatherly

Samuel Glasgow had early sold out his interests in the Centennial Mill, in 1913, mainly because of eye trouble.

He was born October 16, 1858 at La Porte, Indiana. His father owned large herds of cattle and Samuel helped round them up. During the early gold discovery period, Samuel came to Spokane by way of Portland and became associated in flour milling. He was first with the C & C Flour Mills and later one of the founders of the Centennial Flour Mills. He served in Herbert Hoover's Food Administration program. In 1916, he acquired an interest in the Pasco Flour Mill but failing eyesight curtailed many of his business enterprises, although in 1934, surgery corrected his vision. He lived until 1945.

Vanished Gristmills

The *C & C Spokane Flouring Mills* today, with restaurants and shops making it a popular place. It is now called *The Flour Mill*.

The cereal mills in Spokane before the turn of the century, did a brisk business. The old Post gristmill operated with the original C & C as a cereal mill, and the oatmeal mill of the Centennial Mill, operated until it and the flour mill burned in 1909.

An oatmeal mill, said to have been erected in 1887 near the Howard street bridge, called the *Spokane Oatmeal Mill* was 40 X 60 feet in dimension, three stories and a basement, with a daily capacity of 7,500 pounds of rolled oats.

The proprietors were Wadham and Olney, who made a speciality of using only "home grown oats exclusively" to make its products. The mill was destroyed by fire in the early 1900s

Spokane County Gristmills

DYBDALL GRISTMILL, Cheney

In early 1880, two men by the names of Pete Wagner and Henry Hinkey realized there was need for a sawmill in the Cheney region, to supply lumber for the early farmers and homesteaders. They leased some property from the government on the southern tip of Chapman Lake—one of the many charmingly situated lakes throughout the area—and erected a sawmill. A small but strong stream, Rock Creek which came from Chapman Lake ran close by.

A few years later a man by the name of Ole C. Dybdall became interested in this sawmill. Ole Dybdall was a Norwegian immigrant who had recently come to this part of the country from Norway where he had been a sawyer and miller. He was skilled in millwrighting and milling for he had received training and experience in this trade in Norway. Dybdall wanted to start a gristmill in some suitable location, and in 1897 he secured the Chapman Lake property, buying the sawmill equipment from Wagner and Hinkey.

Dybdall kept the sawmill operation going with Wagner remaining about a year as his partner, and in 1897, Dybdall build a gristmill here. At that time, it was known as the Wagner-Dybdall gristmill, located on Rock Creek at Chapman Lake.

At the time Ole built the gristmill, he installed roller equipment in it. By 1897, the roller process for producing flour had taken strong hold and all new flour mills usually put in this equipment.

At first, Dybdall had three sets of the rollers, each a "two-stage" operation. Later in 1910, he added an additional roller cabinet, making four in all. This equipment was supplied and installed by a company in Moline, Illinois, the Barnard and Leas Roller Manufacturing Company. This company also outfitted the mill with the corresponding machinery to go with the rollers, having already put in the plan sifter and the cleaning and tempering equipment. When the fourth roller was added in 1910, the company installed the middlings purifier, the dust-collecting system and added extra elevators, gravity chutes and cupped conveyor belts to augment those installed in 1897. This was now a complete and modern flour-making mill, well able to take care of all the milling needs of the farmers in that region.

Ole C. Dybdall, the elder, operated both the gristmill and the sawmill until 1919, at which time he died at the age of sixty years while returning from an auction. His wife, Annie, and his son Ole C. Dybdall, junior, not quite eighteen years old, took over and continued the operation of the mill. Ole left school and worked

Picture of Dybdall gristmill taken in 1933, when wooden flume was still in place. When turbine not running, water spills over at top. Small building at right houses the sawmill in operation then.
(Courtesy Eastern Washington Historical Society)

steadily in the mill from then on. No one else helped out and Ole and his older sister, Segnie, ten years his senior, handled the milling operations by themselves. Before long, Ole was managing and running the mill as well as his father had done taking care of the ever increasing business brought about by the growth of wheat production on the farms. The charge to the farmer for milling his wheat was thirty cents a bushel. Ole never married. He jokingly remarked, "I was too smart," but there was the mill and his mother and his sister, along with the farm, and then the lake resort that took all his time and attention.

Closeup of system of belted wheels, one set for each roller cabinet.

Spokane County Gristmills

Dybdall gristmill at Chapman Lake, Cheney, Washington. Built in 1897.

Vanished Gristmills

The gristmill building and all its equipment is still standing.* It is idle now although the machinery is still in place, and the old remains of the sawmill repose in the lower yard of the mill building, a jumble of grey weathered beams and boards and a few pieces of rusted iron machinery. The mill building stands overlooking a serene, grassy meadow through which winds a shallow stream bubbling over a rocky bed. Its surface reflects the blue of the summer sky, sometimes mirroring one of the red Hereford animals from the small herd of cattle grazing peacefully on the tender, green grass.

Power for the mill came from water brought from a dirt and rock-constructed dam at the edge of Chapman Lake. The water was carried in a long 500-yard ditch to a wooden box flume and a wooden standpipe. The standpipe brought water to an open turbine-type metal water wheel. This was a small one, thirty-six inches in diameter, one foot in width with seven-inch cups and situated at the lower south side of the building.

In 1940, changes were made in the mill, for much of the early construction was decaying and needing replacement. The dam was rebuilt with concrete, with two large cylinders installed to keep fish from going over the dam. The turbine wheel was replaced with a new one exactly like the old one, and the wooden flume and standpipe were removed and metal ones put up. The new metal flume and the metal standpipe were not conventional pipes in the sense of the word, but an ingenious use of a substitute device serving admirably for this purpose. These were large, steam boilers which were welded together to form a huge, long pipe, about 40 inches in diameter. It was laid at a steep angle beside the south side of the mill leading to the turbine. Here a plate was placed over the "pipe's" end in which was fitted six louvered outlets, or nozzles, through which water shot out to hit the cups of the turbine-wheel in an undershot fashion and turn it rapidly, furnishing the power to the mill.

The metal flume also was constructed by using two or three lengths of these boilers, cut in half with the halves welded together lengthwise. These were placed on a trestle to connect one end of the flume to the ditch and the other to the standpipe.

The trestle itself deserves mentioning, for this was also an invention devised to make use of scrap material. This unusual collection was sturdy and sound. It consisted of several old automobiles, mostly truck chassis frames which Ole "picked up" at old junk car lots in Spokane and brought to the mill site. The frames, all metal, were then placed on end, supporting the metal flume above and with their lover ends embedded in concrete for security. There is no question about the lasting sturdiness of this unusual, and slightly grotesque construction.

*For preservation purposes, listed in National Register of Historic Places.

Metal standpipe and flume. Standpipe is made of 3 or 4 large boilers welded together. Flume made from boilers cut in half and supported on truck chasis frames.

Truck chasis frames support the long metal flume.

This is front of building at which is located the main entrance. As seen, roof is in state of disrepair. Metal flume made from boiler halves appears in foreground.

Turbine wheel at end of metal standpipe. Standpipe's end fitted with six nozzle outlets which shot water to wheel in undershot fashion.

Vanished Gristmills

The weathered building is wooden, four stories high and almost square, each side approximately 30 feet in dimension. The wood for all the mill's construction came from the sawmill which the elder Dybdall ran along with the gristmill. The narrow siding is nailed vertically on the building and the roof is higher on one side than the other, saltbox type and covered with thick shingles. The lower section of the building, the basement, contains the drive room, the floor above—the first floor—housing the rollers, chutes, scales and sacking and bagging station.

The third floor has the plan sifters which worked the flour through sifters in a horizontal motion, and are still in good condition. Here is also the middlings purifier which gave the final step to the flour, separating the bran from it in this operation. The floor above, the fourth floor or attic, contains the bran duster and cleaning equipment. It was to this floor that the wheat was brought by cupped belts after being taken in from the farmer. Here, a cylindrical scourer whirled the wheat at high speed removing the outer husk of the grain. The wheat on this floor, was then tempered. This was a moistening process to prepare the outer covering of bran so it would "hold up" while being taken through the rollers.

The power from the water wheel came to the machinery by means of a series of shafts to which were fitted pulleys and wheels with belts to other wheels, energizing other shafts, on and on. This power originated in the basement from the wheel shafted to the turbine and belted from the basement to a large metal power wheel directly above on the first floor near the south wall. This power wheel energized a long shaft on which were a number of wheels with belts extending to smaller wheels beside the roller cabinets. Here an intricate system of small roller-wheels with belts beside each of the cabinets ran the operations of each of the rollers. Likewise, the entire mill was run with this system of belted rollers and belted metal wheels.

Although Ole said there had never been an accident in his mill, such intricate mazes of heavy, vibrating equipment and many complicated wheels and belts of different widths and lengths, whirling, slapping and moving everywhere at different speeds and rhythms could easily have had a hypnotizing effect. That in itself was dangerous along with the added hazard of these moving parts always ready to snatch at the miller's clothing in an unguarded moment, or catch him unawares should he trip slightly. No miller could ever afford one careless moment. Often, accidents did occur to millers, whether in a mill in Eastern Washington, or in some mill in the eastern part of the United States, many mills in existence there since the 1600s. On more than one miller's gravestone was carved the cryptic words, "Killed in His Mill," which poignantly told the story.

Grain hauled to the Dybdall mill came in gunny sacks drawn by horse team and wagons. It was unloaded at the front of the building with its double doors. The doors

Ole Dybdall, Jr., who ran the Chapman Lake gristmill near Cheney from 1919 until 1950.

were simply wooden planks with a Z-shaped wooden bracing on the back and a metal catch and padlock for security at night.

"Did you lock up in those days, Ole?" the question was asked.

"You bet'cha." He grimly jerked his head in emphasis.

Hand trucks carted the sacks from the wagon to the middle of the room on the first floor where they were first weighed, then dumped down through an opening in the floor covered with a one-inch wire meshing. An elevator below carried the wheat in its cups to the fourth floor to be cleaned and tempered.

The rollers were two-stage, and the corrugated steel rollers in the first cabinet started the process by cracking the grain. Before the second roller had done its separation of this wheat, it already was flour-like. Cupped conveyor belts, enclosed in six or eight-inch wooden shafts, carried the flour back and forth, to and from the sifters on the third floor and finally to the middlings purifier. Ole Dybdall, explaining the working of his mill, estimates that before the final bagging, a barrel of flour went through about fifteen different processes before it became flour. The daily capacity of the Dybdall mill was 60 barrels.

Vanished Gristmills

Dybdall Gristmill built in 1897, operated until 1950.

The Dybdall Mill produced two grades of flour: *First,* fine flour, and *Second,* a slightly coarser grade. The flour was not bleached. The bran was removed in the final sifting and bagged separately. A large, black metal spout in the middle of the room was the station for bagging the finished product. A bag was fitted around the metal frame beneath the spout with flour augered down into it.

When the sack was filled with the flour, the sacker grabbed a threaded sacking needle from a nearby spot, and with a quick flick of his wrist caught one end of the white sack for an "ear", then looped his hand in and out a few fleeting times on the top of the bag, whirled the twine around another pointed ear and deftly snipped the twine with the eye of the needle. The flour sack, now filled, it's labeling standing out boldly against its full belly, had two pointed, white, lynx-like ears which undoubtedly listened and heard all. Like all flour sacks in all gristmills when filled, it now sought its last revenge by becoming dead weight and totally unwieldy, defying all lifting by bulging out in countless unwanted spots, flopping over when it should have remained rigid and pretending to be attached to the floor.

Ole carried on other operations through the years besides the work of the mill.. In 1920, he turned the lakeside property into a resort, putting up one-room cabins and trailers for campers or fishermen who come for the silvers, rainbows and bass. The resort, still going strong, is a popular recreation place for students attending the Eastern Washington State College (now University) in the town of Cheney close by who come for picnics and a few hours' outing. Ole also owned cattle and handled that operation for many years.

Dybdall said the mill stopped running in 1950. Long before then, big mills in large cities had taken over flour production on a wide corporate scale. Small mills could not compete with these operations. Even in 1897, when the Dybdall Mill was built, most other mills also being built then were much larger and carried on more extensive merchant, or business trade, shipping flour to other markets, rather than custom milling, catering to the farmer's milling needs. But it was the production of flour for the farmers that made the Dybdall Mill an outstanding and popular one. Farmers would travel long distances, often one hundred miles or more just to have their grain ground here in the way they wished, "properly", and according to the requirements of the specific brand and grade of wheat the farmer had raised. With the Dybdall rollers each producing two grades in the flour, each finer than the preceding grade, this gave the mill the advantage of an eight-stage milling process, with the Dybdall milling making an exceedingly fine baking flour, the best available anywhere. Most other mills, even the large ones, had six-stage equipment, so there was an extra advantage in having wheat ground at the Dybdall Mill and its process was well known and sought after by people everywhere who brought their grist to the Dybdall Mill for special grinding.

Vanished Gristmills

SPOKANE COUNTY GRISTMILLS

DART'S MILL at Dartford

Around 1879, two young brothers by the names of Lafayette and Herb Dart, traveled west from Mankato, Minnesota, but each coming by a different route. Herb's way led through the Black Hills of South Dakota and finally to the Spokane area. Lafayette's way took him to Walla Walla and from there he bicycled the long rugged distance to Spokane.

They first settled about ten miles north of Spokane at Orchard Prairie, homesteading there. Later, two other brothers, Curt and fourteen-year old George O. Dart also came West joining their older brothers. For a time in the early 1880s, Lafayette and Herb were employed at the new gristmill in Spokane which Frederick Post had recently completed. Here they learned the rudiments of flour milling, and millwrighting to a certain degree. The brothers already possessed a natural ability and inclination towards building and construction, and later they were to use these skills to full advantage.

The brothers took up property not far from Orchard Prairie along the Little Spokane River, nine miles north of Spokane. This location was in a choice, fertile

Sawmill built in 1882 by Herb and Lafayette Dart. Large building at right is livery stable.
Courtesy George T. Dart.

valley with the Little Spokane River steadily coursing through its meadow lands. Before long, Lafayette and Herb built a good-sized sawmill on their property at the edge of the river.

Nature had happily provided an ideal millsite for the brothers by obligingly throwing together several enormous boulders along the river's bank. Some of these were so advantageously placed they could be used as the end of an abutment for a dam the boys built. Later, boulders in another location became a channel for the flume of a gristmill.

A crib dam was constructed, using pieces of lumber, 2 X 12s piled on top of each other with rock and stone placedbehind them for reinforcement. The water in this dam soon collected for some distance back of it, and formed a large lake.

The lumber mill operated successfully, furnishing lumber for the people in the region. Herb put in a small general store to provide goods and necessities for the farmers coming to the sawmill. Lafayette during the years, became interested in buying other properties, which included some lots north of Havermale Island in Spokane, and Herb took on the construction of a gristmill at their homestead on the Little Spokane beside the sawmill already there. He was sure a gristmill would be widely used by the farmers growing wheat up on the Wild Rose and Half Moon prairies.

One of the naturally placed huge bolders on the banks of the Little Spokane River at Dartford, which was utilized in building the dam and flumes for the gristmill, and sawmill.

Herb Dart built the gristmill fairly close to the sawmill and after it was finished, the sawmill and gristmill operated together as a complex, with the exception that the sawmill was operated by steam from boilers, and the gristmill used the water from the Little Spokane collected in the dam. A wide wooden box-flume was built, seven

Vanished Gristmills

Gristmill at Dartford on the Little Spokane River. Powered by three open, horizontal turbines. Water pouring out is overflow from the turbines.
Courtesy Eastern Washington Historical Society.

or eight feet in width, its path following the natural channel so obligingly provided by nature, between the large, almost straight-sided boulders. A few of the edges which protruded out from the sides were hacked and chipped away until they became smoothed and flattened enough to contain the flume.

This wooden flume carried water through and over the construction housing three sets of open turbine water wheels, each placed horizontally. The water flooded down over them and into their cups revolving the wheels rapidly. Shafts extended vertically from the wheels, geared to horizontal shafts which entered the wheel chamber of the mill and energized the equipment there by means of belted wheels and pulleys.

These turbine water wheels provided strong, ample power to run the mill, and two sets (two run) of millstones were installed. One run was a pair of two large granite millstones, each six feet in diameter and probably weighing close to 2,000 pounds. This large run was to handle the great quantities of wheat from the big wagon loads of grain coming in from the farms. The other run was a smaller set of granite stones, each 36 inches across. This run was used for smaller orders or for grinding out corn meal or stock feed when needed, or if the wheat grinding order was a large one, the two sets of stones ground the flour in conjunction with each other.

The gristmill itself was strongly framed to take the weight of these heavy granite stones. Massive timbers and beams from the sawmill were used, braced strongly with joints mortised and tenoned to withstand the constant vibration from the revolving stones and other machinery.

The millstones are said to have been imported from Europe, shipped around the horn to Portland. From Portland they came to Spokane and were hauled from there to the Little Spokane and Dartford by ox-team and wagon. The wagon hauling those stones must have needed extra reinforcement and strong construction to handle even one of those six-foot stones. No doubt, an extra span of oxen was used. In later years, these stones went into oblivion. It is thought that during construction of the new roads, the bulldozer may have moved them and possibly unknowingly destroyed them. One of the smaller stones was kept at George T. Dart's home many years, but he said it disappeared during a period of his absence and was never recovered.

Bolting equipment was installed in the Dart's Mill for the sifting of the flour, with bran one of the by-products which the farmers were glad to have. The best bolting silk usually came from Switzerland and was of different-sized meshings for sifting the flour into fine grades. When the silk on the frames needed replacing, the used material was saved and handed over to the women-folk in the Dart household

Vanished Gristmills

Mr. and Mrs. Herb Dart. Herb built the gristmill at Dartford.

(Courtesy George T. Dart)

Below:
Owners, and mill hands for gristmill and sawmill. At left, Herb Dart and next to him, the bookkeeper, then Lafayette Dart. Names of one or two of the workers are Zeigler and Stoneman.

(Courtesy of George T. Dart.)

who put it to good use. Many a baby's cap was adorned with this glossy, fine-textured material and trimmings were made of it and used for the finer garments of the ladies.

As soon as the gristmill was built the town began developing. No bridges were built then across the little river for people needing to cross it from the north. However, a shallow place in the stream was used for fording. The little settlement thus early took the suitable name of *Dartford.* It is said that Indians from around Colville often came in bands and pitched their tepees along the river's banks on the opposite side of the stream from the town. They would remain a month or longer fishing, the women preserving the fish in their age-old custom of boning and placing them on racks of twigs in the sun. Their tepees formed a colorful colony, bright blankets in the sun, spotted brown and white ponies, and tiny, naked, brown Indian children scampering about in play. Thin lines of smoke spiraling into the sky marked their campfires where they cooked their meals. In the morning after their arrival, the Indians would ford the stream and come into town visiting the stores or bringing their grist to the mill for grinding. They were quiet and orderly Indians and were always welcomed at Dartford.

Original barn and Herb Dart home built in 1883. Barn used for horses hauling for the mill, and for saddle horses. George Dart, III, standing in yard of home, equipped with his fishing gear. The Little Spokane offers many hours of swimming and pleasure for Dartford residents.

Vanished Gristmills

A large red building built at the end of the hill near the river was a store run by Herb Dart. This location became the center of the town. A livery stable was built, another store and a post office was established, a blacksmith shop put up and later a fish hatchery started there. In 1883 or '84, Herb Dart built a large barn for the teams used by the mill, needed for hauling flour or wheat, and for the saddle horses, essential in those days before automobiles. The barn still stands and is in good condition today and still used. A buggy barn for the mill also stood near the horse barn for many years, but it is gone now.

Around 1883, Herb built a large home a few yards from the barn site. This was constructed of wood and lumber from the sawmill. The home is still standing and in excellent condition, the timbers of the foundation said to be as strong and solid now as when first laid there. The house consisted of a basement, a main floor, a second floor and a large cupola on its top.

In the early days, the basement performed a special function, being given over entirely to a kitchen. Here, a good, hired Chinese cook prepared meals which the family enjoyed, as well as a dozen or more mill hands, hired by Herb to work in the gristmill, and sawmill. They took their meals here each day, seated at the long, food-laden table in the center of the big dining room. Cupboards lined the kitchen walls, and pass-through counters stood between the kitchen and dining room for easy passing of great, hot dishes of fluffy mashed potatoes and fresh garden vegetables, and a huge platter or two of roast beef or baked chicken. Outside the house, a "cooling cellar" was dug into the bank of the hill for keeping the milk, butter, eggs and other perishables cool and fresh.

Dartford was, and still is, a little world of its own. In the late summer months, when grasses have yellowed underneath the trees on the hill, the trees and brush themselves remain green and healthy looking everywhere. Clumps of bright green grass line the water's edge, and the air is fresh and moist, delicately scented with the natural perfume of native plants: sweet grass, wild strawberries, service berry, thimbleberries, Oregon Grape and earlier in the season, wild roses, waxy white syringa and blue water iris.

Several members of the Dart family reside in the area. George T. Dart and his wife, Evelyn, live in a beautiful, log-constructed home which blends into the background of fir trees and pines. Antiques such as family furniture of an earlier age, old farm implements along with mortars and pestles once used by Indians and found along the Columbia River, and arrowhead collections are displayed in frames on opposite sides of a great, fieldstone fireplace, Old guns, cross-cut saws and other countless momentoes of the past form a decorative part of the large, beautiful home, both outside and within.

The town of Dartford in the 1890s and early 1900s. (1) Alexander Nelson's store and house, (2) Gristmill, (3) Sawmill, (4) Livery stable, (5) Herb's store, (6) Fish hatchery buildings. Picture is taken from hill looking south.

Courtesy of George T. Dart.

Vanished Gristmills

George T. Dart, whose grandfather George O. Dart was one of the early founders of the town, says that salmon often jumped the dam, and that there is still a good natural supply of fish in the little river. Indians often used to put a net across the mill dam and when the salmon leaped over it, they would fall into the Indians' net. He said that before the Washington Water Power Company put in the Long Lake dam, many salmon came up from the Columbia River to Dartford. George also mentioned that the Little Spokane is unique in that it is one of the few rivers in the United States where a landowner who owns property to the water's edge, also owns to the middle of the stream.

The gristmill operated all through the years from the time of its first being built, never stopping except at night, continuing through World War One. Around 1922, the mill closed down, no longer functioning as a gristmill after that. Business was not so strong as it formerly was, for farmers could now truck their grain to larger mills, selling it outright and buy their flour in sacks of any size at the grocery store. But one of the turbines was wired to a generator to provide electricity for some of the residences and one or two stores. The gearing is said to have made a loud grating sound as it ran which could be heard all over town, and the Dart men were always glad when nine-thirty came each evening, which was the hour when it was turned off.

George Dart sadly recalls that after he returned from the war, the mill was gone. He thinks it was torn down about 1943. Before then, he had taken out the main machinery, but he says the cast iron turbines still lie buried seven or eight feet below the ground near the river's bank. He is sure the wheels would decompose quickly if brought to the surface. Nothing now remains except the site beside the river, a foundation wall of stone and mortar and the beautiful, huge, white boulders.

People living in beautiful homes in the area, carry a deep appreciation of Dartford and its past. No one fords the wide stream now unless they wish to, for near the old crossing a concrete bridge built in 1925, spans the river. Before that date, another bridge was beside it at a much lower level for many years. Its concrete abutments still cling to the edge of the river's bank.

In the 1930s, the mill building was used as a chicken-dinner eating place and was popular, but this did not continue for long. The mill was becoming old and thought not strong enough for further commercial use.

Spokane residents through the past years considered it a treat to visit Glen Tana, a distinctive restaurant for tea or luncheon for an afternoon gathering on the picturesque, peaceful Little Spokane. The acreage on which this beautiful home stood belonged at one time to Lafayette and Herb Dart. The estate of the early Marr family was originally Dart property as well as the acreage bought several years ago by the Kiwanis and now devoted to a Christian Youth's Camp. Herb donated the land for the cemetery which is still used.

The Little Spokane is as strong a stream now as it was a hundred or more years ago and it gives this valley an unmatched beauty. The huge white boulders, placed there so long ago by nature, still carry their significance proudly, for they played an instrumental role in the life of the Dart's Mill.

Store at Dartford run by George T. Dart's father Harold Dart, and grandfather George O. Dart, in the 1920s. **(Courtesy George T. Dart.)**

Vanished Gristmills

SPOKANE COUNTY GRISTMILLS

DEEP CREEK

The Deep Creek locality lies about thirteen miles west of Spokane. At one time during the late 1800s, it was a lively, good-sized town. Today, a few of the remaining buildings still stand beside Highway 2, and passersby in swiftly moving cars no doubt flash interested glances at one or two of them. An old store, Behm's store, is vacant now and boarded up. A "For Sale" sign rests on it and residents say that occasionally someone comes and speculatively looks it over.

The church, and especially the large, old schoolhouse still stands. The schoolhouse is an interesting structure, picturesque and conservatively decorative in its nineteenth century styling.

Farms abound on the wide prairies. They are productive, and the first settlers found that the land needed only a plow to sink into it and turn the sod to ready it for crops, for no pine trees grew there. The trees were plentiful however, in the ravines and on the higher ridges, and still are.

At one time, a strong stream flowed through the region. This was Deep Creek, fed by many springs. It still courses along in its old path but the water in it is not nearly as strong as it was. A picturesque and useful part of the creek was the falls although they no longer exist. Long ago, as the water fell, it collected into a wide pool, forming a natural pond.

Just as the creek and the falls attracted settlers into the region, so it also drew a man who saw in it a potential of power to run a gristmill. This was John N. Bauman, miller and millwright from Portland, Oregon. After talking to several farmers in the region and meeting merchants in the town, all of whom were more than eager for a gristmill, he immediately set to work planning and building one. By late 1880, the gristmill was completed. It had one run, or pair, of granite millstone, each 36 inches across. The mill sat on a small rise near the creek's bank and overlooked the north side of the Deep Creek falls. It took water out of the natural pond, which became the millpond, and the water was channeled to the overshot water wheel by way of a flume, pouring down on the wheel from overhead, to turn it.

Business at the mill became so strong that before long, Mr. Bauman decided to change to the roller process, which all gristmills then were beginning to use. He then replaced the water wheel with a turbine to step up the power and the output of the mill increased from 50 barrels a day to 75 barrels. He called the mill, *"J.B. Bauman and Sons Flour Mill"*.

Mr. Bauman also ran an eating place, mainly for travelers along the White Bluffs wagon road.

At one time, the stage line from Walla Walla to Colville passed near Deep Creek town, crossing Coulee Creek which was about four miles west of Deep Creek. It is said that the wagon ruts are still visible in many places. This crossing has some historical significance, for it is about one-half mile east of a location where Isaacs I. Stevens, the first governor of Washington Territory, and Captain George B. McClellan, later General McClellan of the Civil War, camped for five days in the fall of 1853. They had been entrusted by the U.S. Government to locate a northern route for a transcontinental railroad. There is a marker on Coulee Hite road, west of Woods Road commemorating this meeting.

Besides the gristmill which brought a large amount of business into the town, a cheese factory was built along the banks of the creek, which operated for several years. Later, it was bought out by the Hazelwood Dairy of Spokane who kept it going until it went out of business itself.

Few people are still living who remember the gristmill in Deep Creek. Many recall that it was an important one and that their grandfather and all his neighbors used it. The first men to take up homestead claims here were two brothers by the names of Daniel and Alfred Stroup who came in 1878. In a few years their farms were producing good crops of wheat and they became steady customers of Mr. Bauman's mill. Henry Eads came around 1878 also, and he put up a little building and stocked it with canned goods and a few staples such as sugar, and coffee. The Louis Magers and the McDowell families came in 1878, arriving in late November scarcely a minute too soon before winter clamped an icy hand over everything. They are said to have literally "holed in" for the long cold season by digging into a hill slope, throwing the dirt up and over their abode for protection from the weather, and laying logs overhead for a roof. This was only a tiny room, but both families lived in it, thirteen of them, grateful for each other's company and protection. In the coming spring of 1879, they filed on their homesteads, and before long were raising their wheat crops.

A daughter of Louis Magers, Mrs. Mary Flood, still lives on the old homestead with her daughter, Delores Fels and her family. Mrs. Flood is ninety-four, having been born in 1883 in Deep Creek, and in 1977 she proudly said, "I'm the oldest person in Spokane County, who was born in the county".

She remembers that her father took grist to the mill, but she says she never visited it herself. She remembers the stores and the town and the people and the time when food was hard to come by, and they many times ate only boiled wheat for supper, with milk from the family cow. By 1890, a number of stores were operating and about 300 people lived in Deep Creek. Besides the flour mill and the cheese factory, there were two hardware stores, a specialty shop operated by Mrs. Young, two general stores with one run by Thomas Perkins and the other by M.S. Gardner, a grocery operated by A.M. Merriman, one drug store, two blacksmith shops, one run by Frank Perkins, two hotels, two livery stables, a butcher shop, shoe shop, a sawmill and a schoolhouse.

She recalled a few names of the first people living in the area; the Gilstraps, McDaniels, Douglasses, Harringtons, Alf Thomsons, Bill Traynors. There were the Hyslops, the Rombecks, the Hotchkins family, the Greenes, the McFarlands, and many others. She said that before the post office was established in Deep Creek everyone had to go to Colfax to the post office there. "My father often walked to Colfax from Deep Creek to get his mail," she said.

Dr. Waterhouse was the local physician, as well as dentist in a pinch, and a farmer running his farm on the side. Mrs. Flood remembers he pulled a tooth for her when she was eight or nine. She was promised a doll if she would behave nicely and let Dr. Waterhouse take out her tooth. "I got my doll," she smiled. "We bought it at Mrs. Young's store which had a lot of interesting things—laces, ribbons, flower trimmed bonnets, dress materials of pretty calicos and silky poplins, and all kinds of dolls. Mine was a big one, and oh, how I loved that doll for it was my very first one."

The gristmill's business was said to have been tapering off by 1890. By now, Spokane was easier to reach than it was in the old days and instead of the old gristmill which Frederick Post had built in 1878, several others were standing—big, efficient flour mills which turned out excellent white flour.

Schoolhouse at Deep Creek where students of all the grades and high school attended. School is not used now.

The depression of 1893 was hard on folks everywhere, especially on farms. Many properties were foreclosed, the land changing hands and the older settlers giving up their farms and moving elsewhere to find another way to make a living. By 1895, it is hinted that a few remaining found another way of staying ahead. A rash of fires broke out, and it is thought that some might have been tempted by the insurance which they collected, but there probably never was any real proof of this. Whether the gristmill suffered the fate of fire, or was later torn down after it stopped running in the early 1890s, has not been determined as of this writing.

There is nothing visible to indicate a gristmill stood beside Deep Creek and near the falls, except some concrete footings, and what may have been a ditch. But they must be hunted out, for grass and brush hide everything of that nature. The Deep Creek area is still as attractive as of old, and the few old buildings remaining help it to cling to the memory of its better days, when the gristmill was an incentive for farmers and their families to come to town and shop, while their grist was being ground at Bauman's mill.

SPOKANE COUNTY GRISTMILLS

MARSHALL

Marshall is a little settlement eight miles southwest of Spokane, for some time mainly known as Marshall Junction. No town exists there now but in the early 1880s, or before, it developed into a small one.

William A. Marshall came there in 1878 from California. He particularly liked the location of this little site (which later took his name), because of the possibility of water power. He homesteaded the west quarter of section 22 on which was a strong spring and from which flowed a stream. Since Mr. Marshall was a lumberman by trade, he decided to put in a sawmill using the water for power. He made a dam and put in an overshot wheel to run the sawmill, put in an up and-down saw and established the first sawmill in that locality. It operated for several years producing a limited amount of lumber.

In 1886, Marshall sold all his interests, sawmill and property, to W.E. Pierce and George A. Davis, These men were experienced mill men, sawmill and gristmill, and came originally from the State of Maine. In these years, the Spokane and Palouse Railroad was built, and at Marshall the railroad branched off from the main Northern Pacific line establishing a junction at this place. Pierce and Davis platted the town of Marshall in 1887 and began improvements in the old sawmill as well as making plans for a gristmill.

They took out the up and down saw and put in a circular one and they built a gristmill. The gristmill also used the power from the stream, and the two mills probably worked in conjunction with each other, with the same overshot waterwheel supplying the power for both.

In 1888, George Davis bought out Pierce and became sole owner of the mill and the townsite. Mr. Davis built a new sawmill which used steam power and had a capacity of 60,000 feet of lumber a day. The little town began to boom and a lot of relatives and friends of Mr. Davis and Mr. Pierce in Maine moved out to Marshall and started taking up homesites.

The gristmill with the water rights and land adjacent to it was sold to a Nelson Martin of Spokane who was associated with Mayor Fotheringham, it is said. He put in up to date machinery, roller equipment, and operated the gristmill under the name of *Spokane and Marshall Milling Company* for about a year.

The gristmill then lay idle until 1898 when it was purchased by W.H. Wiscombe and W.E. Pierce. The mill, operating under the name of *Superb Rolling Mill*, turned out good grades of flour. The Lynchs were long-time residents who farmed in the region. Another family, the McKenzies, were also early residents and Ross McKenzie married Frances Lynch, daughter of Frank, and they ran a dairy at Marshall for many years. At one time in the early 1900s, Mr. Amery owned the flour milling company with Frank Lynch the miller.

The gristmill then lay idle for several years until 1898 when it was purchased by W.H. Wiscombe and W.E. Pierce—the same W.E. Pierce who originally had been associated with the gristmill with Davis. The mill started grinding wheat again and turning out good grades of flour, operating under the name of *Superb Rolling Mill,* and turning out 100 barrels of flour a day. A number of important improvements were made by the new owners with the installation of an improved cleaning apparatus and bolting equipment to sift the flour to finer grades, with bran a by-product.

Marshall town has always been an important railroad-switching town. Railroad tracks there still groan under long, heavy freight trains. A few weathered shacks stagger weakly along the creek bank. They once belonged to railroad maintenance men but now are only grey ghosts reaching for the cool waters. Here at Marshall the Northern Pacific railroad branched down into the Palouse country, and along with the Milwaukee and the SP&S, (Spokane, Portland and Seattle), switched for clear passage to other points.

The town proper, once an active village, was, and still is located over the bluff-like rise beside the tracks. The Frank Parks family ran the hotel, also the combination grocery store and post office, with Frank working at railroad maintenance part of the time. Another store was run by the Filiatreau family, and in the 1930s, John Edmonston owned and ran the hotel-store. These no longer exist, but a little store with an ancient look about it still operates at Marshall.

In 1904, a small factory to manufacture cement blocks was built there by the Oriard family and around 1910, Leon Oriard, senior, started the dynamite factory, known as the *Oriard Powder Company*. It was located on the hill west of the creek. The dynamite was shipped from San Francisco by boat to a site near Lacey, Washington, and stored there. The Oriards brought loads of it back to the Marshall magazine where it was manufactured and boxed. Many sadly remember a tragic accident that occurred in June, 1948, when Leon, a brother of Marcel, was killed in an explosion.

At one time, an economic improvement evolved with the use of the explosive material changed from dynamite to fertilizer, this discovered when a load of fertilizer in another part of the country "blew up." The Marcel Oriard's son, Hugh, has a patent on the resulting new explosive product.

The Oriard Powder Company now operates at Deer Park instead of Marshall. The buildings at Marshall were torn down several years ago.

A tranquil beauty prevails in the Marshall district. Railroad tracks still trace long pathways through the area, but marks of other industries; the gristmill, the cement factory, the dynamite plant, and even a greenhouse which flourished for a time, are all erased. Many newcomers do not realize a gristmill once was an important part of their town, but older families know and keep the memory alive.

Vanished Gristmills

SPOKANE COUNTY GRISTMILLS

JOE HENRY'S MILL

One of the most enterprising businessmen in the southeastern part of Spokane County in the late 1880s and early 1900s, was a gentleman by the name of Joseph Heinrich. Over the years, beginning sometime before 1890, he constructed and operated three mills, two of them were sawmills, and the third one was a gristmill.

Mr. Heinrich was seldom called by that name, for everyone who knew him or his mills called him Joe Henry.

Joe Henry's Gristmill on Hangman Creek. The elevated wooden box flume shows at left, and the spokes of the water wheel are visible at end of flume. Picture 1890s. **Courtesy Glen Adams**

roperty on Rock Creek. There is more than one Rock Creek . The Rock Creek here is a stream forty miles in length with l tributary, and running into Hangman Creek. This creek, originally called Latah Creek, before George Wright's dians at a point along its banks.

ill was a small sawmill on Rock Creek on his own homestead lenty of pine trees around the area to keep his saw busy and a :d among the farmers and settlers for his planks and boards. l called the Sylvan School, used lumber from his mill.

went into disuse, needing new machinery, Joe Henry put up v one at the little settlement of *Duncan* not too far from his illsite. This operated with good results, for Henry knew his t good grades of lumber.

many and had received a certain amount of training and llwrighting and milling. He had long thought about building l off because the farms were scattered and sparse when he first came there. Now, more settlers lived in the area and crops were good. The only other gristmills close by were the ones at Post Falls, or Pine City, long distances away, so he knew he'd have the business of the farmers if he had a gristmill operating.

In 1890, he built his third mill, the gristmill near Smyth's Ford. This one was on Hangman Creek, and described by people who live near there as "down the hill below Fred Felgenhauser's (G. McNeal) place on Hangman Creek."

This location on the creek was unfortunately situated in a canyon, rather difficult to reach by the wagon roads, little better than trails used in those days. But it was a good place otherwise for a mill, for there was a level spot for it and plenty of water available from the Hangman Creek to run it. Green trees and palisade rocks forming columns on the hill opposite the mill, made the location a picturesque one.

Joe dammed up water from the creek into a pond, and from the pond a ditch brought water to the millsite. There a wooden box-flume was built on supports high enough to bring the water directly to the top of the overshot water wheel. The water poured down into its pine cups and turned it, running the machinery of the mill. The main equipment of the mill was the one run, or pair, of millstones.

Most millstones, which were often also called burrstones or buhrstones, were fashioned of granite, but occasionally, although not often, they were constructed of iron, and it is said that Joe Henry's millstones were of this material. No information as to their size is available, but they probably were at least two feet, and possibly three, in diameter.

Joseph built all the mills himself, with all the workmanship done by hand, using the tools of those days for chipping smooth a big timber's length, usually the adze and broad axe. Smaller tools fashioned the tenon, or tongue of one beam mortised snugly into the properly shaped hollow of another. An eight or nine-inch length of wood, usually hard wood, was whittled into a "pin", even to its round head, and tapped into the mortised joint to secure it even more firmly.

The gristmill built by Joe Henry, was a sturdy, three-story structure. Long, narrow boards, one by twelve feet, were used for the siding, put on vertically, and the building itself was well able to hold several loads of filled wheat sacks, or the manufactured flour which was ground out from between the millstones. It was well able also, to withstand the vibrations and weight of the two large iron millstones.

No nails were used in the interior of Joe Henry's mill, but square nails may have held the siding to the framework.

Much of Henry's mill machinery was fashioned of wood. The big power wheel at the end of the shaft of the water wheel, was wooden. This could have been belted to another wooden wheel, or if cogged, meshed with the cogs of another smaller wheel. Often the cogs of some wheels were wooden also, mortised into the side edge of the wheel, although this may not have been the case in Henry's mill. Joe would have made the overshot wooden water wheel, and wooden box-flume carrying the water to it.

One of the early problems of Joe Henry's mill was its location. A farmer had enough headaches without the added possibility of his wheat sacks being spilled out of the wagon, or even the wagon tipped over itself going the last few hundred yards to a mill with his grain. As his wagon started down the bumpy dirt road to Henry's mill, the farmer would brace his foot against the old shoe brake and pull back tightly on the horses' reins. But instead of this slowing down the old buckboard, it more than likely turned it into a sled which began skidding giddily over the rocky, washboard road. The horses would jerk up their heads and their eyes would flare wildly, and the farmer would tense up.

By the time he had reached the bottom, the farmer's blood pressure was elevated just enough for his temper to be sparked off easily. And it is said that Joe Henry could fire a temper easily for he had one himself, and if not irking a customer with his crustiness, could always manage to say the wrong thing and begin an argument. But he also could be affable with rough, good humor many times, too.

He could tell many an exciting tale to the younger boys who visited with him when their father let them go to the mill with him. Bill and Harold Bare who lived above the mill, remembered that when they were young boys, they skidded down the

Spokane County Gristmills

Joe Henry's gristmill on Hangman Creek in the 1890s and 1900s. The man sitting in foreground is Joe Henry, whose proper name was Joseph Heinrich.

Courtesy Eastern Washington Historical Society.

hillside on the backs of their heels and ended up at the mill talking to Heinrich. He would tell about the years he had spent in New Zealand and Australia and about his own two boys.

According to the *Fairfield Community Study,* a history of the area, "One of Henry's sons was a professional wrestler when the family lived in Fairfield, Washington, and in 1915, was admired for his exploits in that field, ...pounding the mats", old Heinrich said, "with Shorty Pottratz, George Doepke and William Reigel."

Joe Henry's disposition took on a darker, menacing tone during the World War One days. He was pro-German to the core and entertained not the slightest doubt or thought but that Germany would win the war. He would not give an inch when talk turned to the Germans, and the Kaizer, and the "Huns"—the worst word anyone could be called. He fought the war all by himself for Germany right down at his mill—Joe Henry against the entire enemy countryside, and he became unpopular. His business fell off and people around the area avoided him and took their grain to other mills which had recently been built closer to the area.

Joe Henry is said to have died shortly after the war ended in 1918. He was up in years then, tired of working, and besides, he and Germany had lost the war.

The mill closed for a time after Joe Henry's death, but later one of his sons opened it up and started running it again. The mill was said to have put in the roller process, a much more modern innovation than the millstones. Still it did not have all the equipment that the other more up-to-date mills were using to produce a white, bleached flour. Even the mill at Rockford, less than ten miles away, by now was producing a fine grade of flour. It had recently been thoroughly overhauled with new equipment added, so it was only natural that farmers took their grain there to be made into flour. Soon the old Joe Henry mill fell into disuse, suffering from lack of business and was closed down.

The mill was finally torn down in later years, after it had stood idle a long time. The old planks and beams in the mill were still as sound then as when first put there many years ago by Joe Henry.

Spokane County Gristmills

CHENEY

John C. Davenport came to Cheney from Colfax in 1880, and built a flouring mill in Cheney at that time. He also operated a store in Cheney and ran a bank, managing these businesses until 1884, when he moved to Coeur d'Alene to become involved in mining. John C. Davenport built the first gristmill in Colfax, in 1873.[1] The first Cheney flour mill was operated with steam.

In 1890, the Cummins Brothers and Company took over the gristmill at Cheney. This company consisted of J.F. Cummin, C.F. Cummin, D.N. Fish and Dr. F.A. Pomeroy. They equipped the mill with the roller process machinery and operated it under the name, *Cheney Roller Mills.* In 1897, the Cummins Mill Company doubled the capacity of the mill. On April 5, 1898, the Spokesman-Review noted that the Cummins Mills received an order from Australia for 10,000 barrels of flour.

During this period, Frank Martin, who farmed property between Cheney and Medical Lake, became associated with the Cheney Roller Mills. When the Milwaukee Railroad was putting its track through Cheney, the mill company put up two large hay barns and Frank Martin sold hay to the railroad company for the horses used in grading the track beds and putting in the railroad. One of the hay barns, used as a warehouse, is still standing.

In 1915, the old wooden mill building took fire and the *Cheney Roller Mills* burned to the ground.

* * *

In 1918, Frank Martin put up a new roller mill to replace the first, old one. Pete A. Wagner is said by some sources to have been associated with him in building the mill. It was a large, four-story concrete building with roller equipment, and powered by electricity which the Washington Water Power Company brought in "on poles" from Spokane. The Northern Pacific railroad laid a track beside the building for easy loading of the sacks of flour. These sacks all bore the unique trademark of the flour mill—a large, red apple.

Frank Martin, born in 1883, ran the mill with great success. It was updated and improved periodically. He took his son, Clarence Daniel Martin into business with him, and Clarence ran the mill until it was sold to the National Biscuit Company,

1 *History of Spokane County*, Edwards

June 3, 1943. This company still owns it, buying wheat from the farmers, processing it and shipping it to Portland.

Clarence Martin had an illustrious career other than being a successful business man. He became governor of the State of Washington in 1931 and served two terms. After that, he returned to Cheney to make his home there, passing away August 11, 1955.

Spokane County Gristmills

ROCKFORD

A flour mill was built in Rockford in 1889, a year after the town was incorporated. As early as 1879 a sawmill was put in using the water of Rock Creek to run it, and built by Farnsworth and Worley.

The flour mill was a good sized building built by the Sheldon brothers. The roller process machinery was installed, with other apparatus of cleaning and bolting put in to go with this equipment. Whether the mill was powered by steam or by turbine, using the water from Rock Creek is not known. It is very likely that turbines were used for the output of the mill by 1891 was 800 *bushels* per day, according to the Spokesman-Review, August 26, 1891.

In 1897, the mill, called the *Valley Roller Mills* was thoroughly overhauled and made "practically into a new mill." The latest equipment available was installed in the mill, the demand for its products developing into a wide market. The capacity of the mill now was 300 *barrels* per day and the brands were "Fancy Patent", "Klondike", "White Star", "Superfine", and "Oriental", this flour going to Seattle, Portland, San Francisco, eastern cities as well as to Hong Kong, China.

This mill, like many others of a similar nature by 1900 no longer had enough business to remain open. No information is available as to whether the mill burned or was torn down, but nothing remains of this structure now.

During the 1930s and 40s, another mill operated in Rockford. It was a small mill in the "middle of town", and called a *Little Marvel Mill.* This produced a satisfactory grade of flour and many farmers brought their wheat here for grinding. It bolted its flour, separating the bran and shorts which Mr. Cosman, the operator of the mill, gave the farmers for their stock feed.

Today there are large wheat elevators in Rockford and Rock Creek still flows through the area, but flour mills here long ago became extinct. Farms produce abundant wheat on the rolling, picturesque acreages, as well as bluegrass, and the yields average high.

It is said that the wheat grown on most Palouse farms in the 1890s and early 1900s was more a pastry type of flour, rather than a bread flour for it did not contain enough gluten for bread flour.

In later years, varieties were planted which had more gluten content, and these were the red and semi-hard grains such as Triplet wheat. Bluestem and Marquis wheat, the spring varieties, were also high in gluten content.

Vanished Gristmills

COX'S FLOUR MILL in Spokane

The Cox's Flour Mill was built at Cox's Park, a crossroads. The location of the crossroads was just north of Hillyard and a mile south of Mead, and in the early 1900s, was called Stockman's Park. The East road led to Peone Prairie and Mt. Spokane, and the West road went to Five Mile and Dartford. Mr. Cox was a relative of the original owner and the crossroads later took the name of *Cox's Park.*

About 1910 or 1912, a man by the name of John Anderson built a flour mill on the northeast corner of the park. Later it was purchased by Frank McGee and his uncle who relocated the mill on the northeast corner. Frank McGee had a serious accident in the mill, becoming entangled in the maze of belts of the machinery. He was so badly injured the family decided to sell the mill, and in 1922, William Whipps bought the property.

The mill was again sold in 1924 to Charlie Graham and his son, James. Within a short time the mill was running at full capacity producing a hard wheat flour of high gluten content.

The mill was not water-powered, and probably was operated by electricity, but it ground flour with a run of granite millstones. It is included here because it was a popular, early Spokane County flour mill which did not convert to the roller process

The mill had a capacity of 50 barrels of flour a day "when run in three shifts", and the brand of the flour was known as "Inland Special". Graham, the owner, also ground stock feed for the farmers if they needed it.

The mill was sold in the 1930s to the Boyd-Conlee Company. They still own the land although the mill was torn down shortly after they acquired it. A grain elevator stands on the property which stores the wheat raised by farmers in the Peone Prairie district

Feed and Flour Mill at Cox's Park north of Spokane

Vanished Gristmills

Garfield County Gristmills

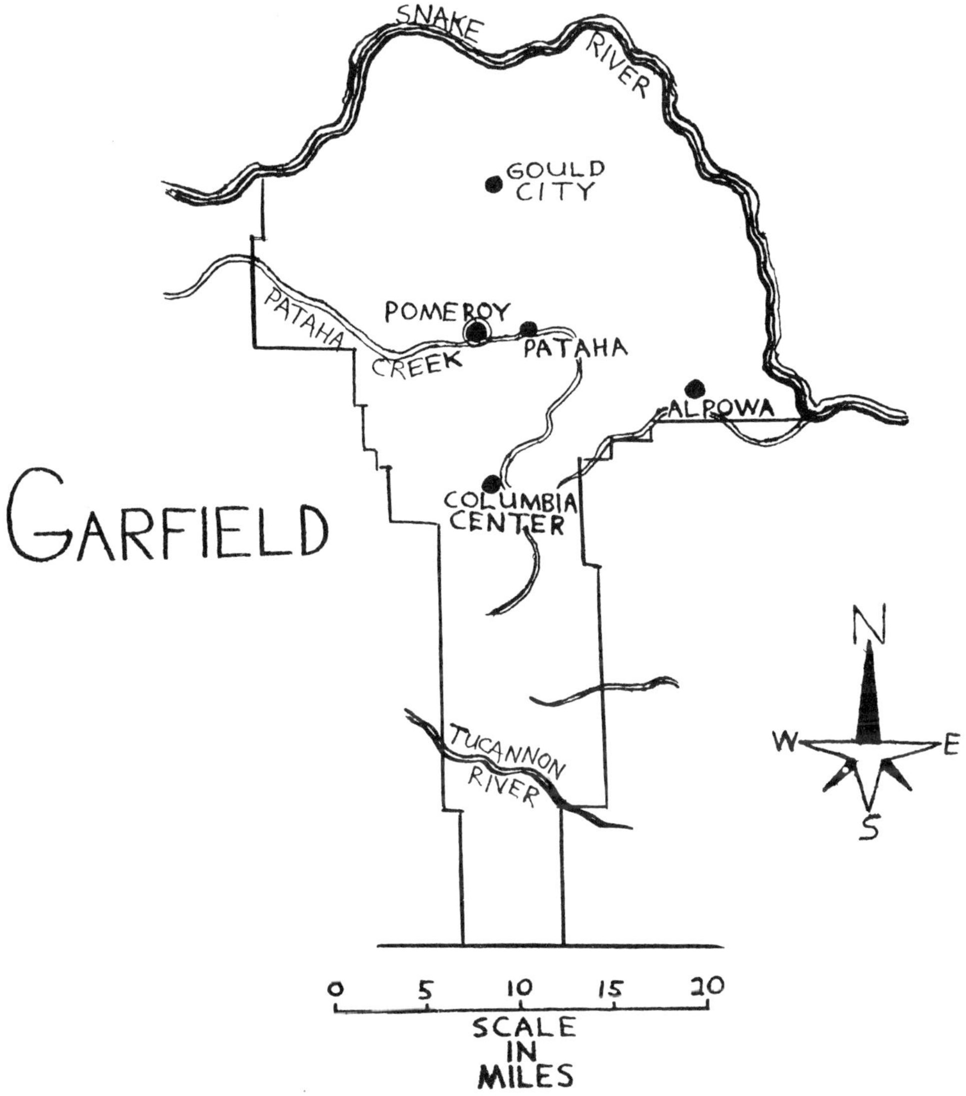

GRISTMILLS IN GARFIELD COUNTY

Columbia Center
Pomeroy Flour Mills
Pataha City Flour Mills
Alpowa Gristmill
Gould City Mill

Vanished Gristmills

GARFIELD COUNTY GRISTMILLS

COLUMBIA CENTER GRISTMILL

A WIDE band of ultramarine blue borders the southern regions of Garfield County, its uneven edge sharply outlined against the pale horizon. This border is a section of the Blue Mountain range, and so close to the town of Pomeroy that only a half hour's drive leads to its foothills and deep canyons. Here many steep slopes, blanketed with fir and pine, catapult downwards hundreds of feet, to pause a moment beside a cool stream before leaping up again on the opposite side. Each canyon is different. Some are so dense with fir trees, the area resembles a wilderness where only deer or other wild animals could live. Other hills spread out more widely to form graceful valleys. Often the valleys are roomy and flat enough to become the site of a small village nestling close to a stream rapidly threading its way along the valley floor.

Columbia Center was such a little town near the Blue Mountains. Its stream was the Pataha Creek, where it forked into the Dry Pataha. The Pataha Creek still flows through the area, hesitating only long enough to make a sharp bend in the center of the valley before it tumbles along over its pebbly bed to the town of Pomeroy about ten miles away.

At first, all this area was Columbia County, with Dayton the county seat. The new town was in the geographical center of Columbia County, so it chose *Columbia Center* for its name. For a time other towns were given the opportunity to compete with Dayton as a new county seat, and Columbia Center carried a secret hope it might be selected, although this did not occur.

Thomas G. Bean and Andrew Blackman were two men most enthusiastic about starting Columbia Center. They felt it held great potential as a farming village and their immediate initial plans for it was a gristmill, which the farmers in the area badly needed. Bean and Blackman already owned a sawmill with a circular saw,

Garfield County Gristmills

Columbia Center gristmill near Pomeroy. First location was in foreground. Gristmill building now used as barn.

which could supply lumber. This sawmill replaced an earlier whipsaw mill on the creek operated by Henry Sharpnack in 1873.

These two men obtained the property for the townsite from John Wentworth who had taken over 160 acres earlier. They bought it from him in 1876 for the "$100 Homestead Cash Entry", and in that year, 1876, they immediately platted the town and began formulating plans for the gristmill.

Although the purchase from Wentworth of the property was in 1876, the filing of the record at Dayton, some distance away, was not made until a year later, December 26, 1877. For the purpose of this story, only a portion of the purchased property is of main interest and that is a small part of the township, lots 15 and 16 of block 2. This was the property on the creek where it made its bend, and this was the gristmill site.

The two men now had their gristmill location and the promises of cooperation from the farmers to bring in their wheat to keep the mill operating, after it was built. They were sure, also, they had the right man to build the mill, but first, money was needed. Casper (Cass) Corus, who "owned the property on the hill" agreed to put up the cash necessary to get the mill going, and he also took the responsibility of putting in the ditch and the reservoir to supply the water. For the millwright, the two men secured the services of a skilled millwright and miller, Luther A. Stimson, whose fine reputation in the milling line had already preceded him to this region.

Luther Stimson was originally from Des Moines, Iowa, and he and his wife and family of children had recently moved to Anatone, near Asotin. Luther came to the Snake River country well recommended, holding affidavits of recommendation from owners of many mills he had built before coming West.

The gristmill was soon under way and the site was a busy place. Lumber was hauled in, the sound of the ax and the adze, chopping at big timbers, could be heard all day long, hand saws grated and screeched as men worked them through thick, stubborn pine boards, and hammers pounded constantly. Huge beams and boards were soon in place making the building strong enough to handle heavy granite millstones and massive machinery and take vibrations from huge shafts and wheels. The ditch, which started from the dam taking water from Pataha Creek, was a long one coming along the hillside "above the Peola road", ending at an incline above the mill. From there a wooden box flume, about 15 or 20 feet long, rested on the natural slope of the incline and carried the water down to the water wheel, in an "undershot" fashion.

There was a sluice gate constructed in the dam, which would release some of the water if it became too high, especially during high-water seasons. There also was

The *nether,* or bottom millstone of Columbia Center gristmill, cut and fitted in sections. *Furrows,* grooves, radiate out from center. Wide iron band encircles stones. The top *runner* stone was said to have been crushed.

Bracing of pine beams with a mortised and tenoned joint, pinned with a long, stout wooden peg with a round head.

Notice the cuts in upper beam. Boards were placed in these cuts and planks laid on top of them for a floor.

Also, note the graceful shaping of the post into a pillar-like support.

a control sluice at the end of the box-flume, near the wheel so the stream coming to the cups of the water wheel would be about six inches deep, just deep enough to keep the wheel moving steadily—"Without shock," according to Stimson, and run off afterwards in a tail race so it wouldn't back up and swamp the wheel. The water wheel, which Mr. Stimson built, was 8 to 10 feet in diameter. It was said to take about three revolutions of the water wheel to make one turn of the millstones.

The millstones, which were purchased from a firm in North Carolina, were 36 inches in diameter, composed of a "pebbly quartsite" which was a "hard type of granite." The top *runner* stone was fitted with a metal device in its center, or eye, into which locked the shaft which turned it. It was less thick than the bottom, or *nether* stone. The nether stone was in nine fitted sections like a jigsaw puzzle and was held together by a thick, iron band eleven inches wide.

The mill was finished early in the year 1878, ready to start grinding flour. It was a three-story structure, 26 X 36 feet in dimensions. Its first floor was made of stone and mortar two or three feet thick on which rested the two top floors of wooden construction. The inside was strongly framed with ten by ten beams, stoutly braced with carefully fitted mortised joints which were tenoned and pinned. Mortising was the hollowing out of an opening in the side of a beam into which the shaped tenon, or tongue of the supporting beam was fitted. This resulting union was held together with a long, tapered wooden peg inserted through them. The big beams were hand hewn and no nails were used in the interior of the mill.

Thick, long floor joists had cuts the size of planks which were placed in them to form a base on which heavy, wide floor boards were laid. The narrow cut siding on the outside of the building consisted of boards of different widths and lengths just as they came from the old sawmills, and were secured with square, iron nails.

In this mill, built on simple plans, there was just one run, or pair, of granite millstones. These were placed in the most stoutly braced part of the first floor, for they were heavy and though put together to be perfectly balanced, they still vibrated and shook even the strongest room. The flour from the stones was caught at the sides, captured by a wooden enclosure around the stones. Here it was scooped up or

Mortised and tenoned wooden joint, pinned with a stout, long, wooden peg with a rounded head. Beams are ten by ten inches, and pine.

shoveled into a bin close by for storage, or placed into the waiting farmer's white, homespun sack. The sack was then sent down below, via the smooth, wooden chute near the millstones where it could more easily be picked up for carrying away.

The mill was a comfortable-enough place in the warmer months of the year, but the winters would find it unheated and cold, work stopping early in the afternoon in the dark, winter months. Danger of fire kept a stove or fireplace from being put in the mill, and even candlelight was scarcely every used. The buildup of flour dust always in the air, could so easily cause combustion that any spark or flame was avoided. Even wooden scoops or shovels were used instead of metal ones to lessen the danger from sparks.

Farmers were moving rapidly into the region now, and by 1878, a number of farms on the high ridges, the prairie and the flat were producing good crops of wheat.

Other bracings in the mill and large beams, all mortised, with pins of wood. No nails were used in the interior of the old mill building which is still standing.

The Humphreys were here, also the Harris and Rush families. The Baldwins were already in the region, having come in 1875. The three young Baldwin boys, Charles, Zenus and Noah Oscar (always called Oscar), would grow up here.

W.F. Cluster, L.B. Courtney, W.F. Shawley, Ranson Long, Rev. Calloway, Derrick Zemmel, Robert Story, Wm. Queen, the Stallcops, and the Rose family who came here first, all used the gristmill. Other farmers from as far away as Anatone and Asotin, thirty or forty miles distant, brought their wheat to the mill.

Luther Stimson helped get the mill under way by running and operating it for a time until the community was able to find a trained miller who would handle it.

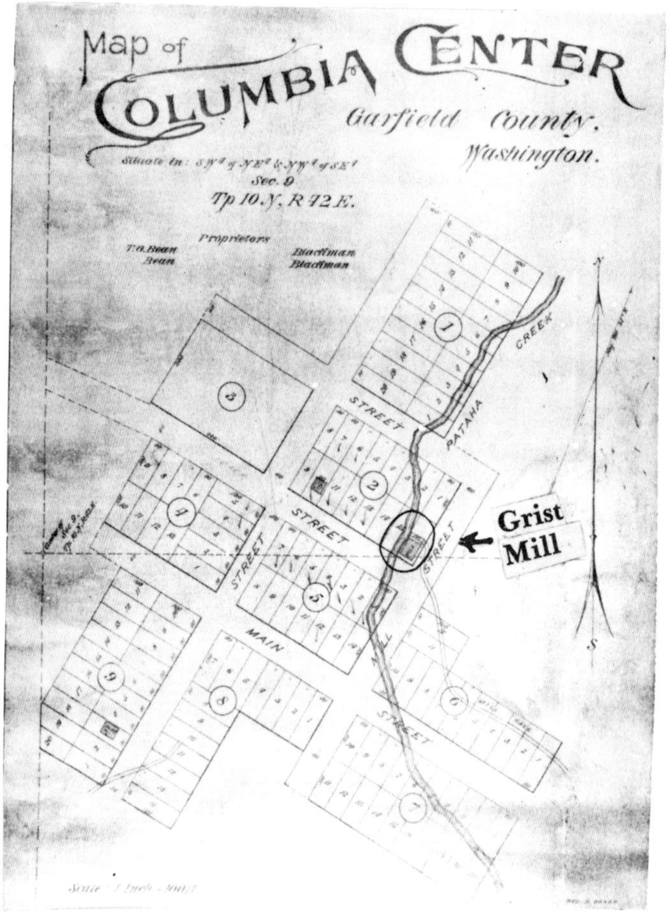

As soon as the gristmill was in operation, other buildings started going up and businesses were opening at Columbia Center. The post office was taken over by J.B. Hastings, who also had a little store in the same building. Hastings was a preacher and a Justice of the Peace as well. There were two livery stables, two blacksmith shops, with one operated by Matt Hudson and the other by James O'Connor. Drew ran the restaurant and there were two saloons, one operated by Dick Travis and one by a man named Backenstoe, (who originally was from a circus) acting as helper. One meat shop was owned by the King brothers and a shoe repair shop was operated by Jim Gallagher who was a son-in-law of Reverend Eliel Oliver.

A general merchandise store was run by the Hutchens family and there were several other stores and restaurants. A schoolhouse was built very early which was a small, one-room cabin of rough boards. There were no churches yet, but Reverend

Garfield County Gristmills

View of Columbia Center town from highway coming into town from Pomeroy. Gristmill is larger building in lower left corner. Picture taken in early 1880s.
(Courtesy Garfield County Historical Society.)

Hastings held services on Sunday mornings wherever there was a place big enough. If a lawyer was needed, one sometimes came in from Dayton or Walla Walla, and a doctor nearby took care of the sick in the region. Nearly 500 people lived in the little community which was becoming a very busy place. Many folks running the stores, also farmed.

A few months after the mill started running, a miller was engaged. His name was Blonk Wennery and he came from Missouri. He had run mills in Arkansas and Missouri, and was married. The young couple must have considered this a fine opportunity for them in this flourishing western town, with shops and stores and friendly neighbors. There was something always going on.

Under Wennery's operation, the mill ran every day from morning until forced to stop by darkness. Farmers bringing in their wheat often left it at the mill for Wennery to grind while they went across the creek to trade. There was usually a piece of harness needing mending, or maybe a list given him by his wife which had to have attention, "Pickup spool of white thread at Hastings and get a couple of 2-cent stamps at P.O." It was always good to visit with another friend who was in town like himself to 'go to mill' and do an errand or two. Weather always got its share of the conversation, and, of course, politics.

Vanished Gristmills

The town's citizenry were somewhat surprised when the gristmill's ownership changed hands in the spring of 1878. The records noted, "In May 1878, Andrew Blackman sold his half ownership of the mill property, lots 15-16, block 2, and other Columbia Center property to A.B. Bean, for $2000."

Shortly after this on April 20, 1879, T.G. Bean, the other original owner, sold his half interest to one of his relatives, possibly a son, A.B. Bean, for $2000. This was the mill property, lots 15-16, block 2, along with other lots.

Andrew Blackman and T.G. Bean were no longer owners of the mill which was now under the sole ownership of A.B. Bean. The mill, however, functioned as before.

The year was now 1880, and the little community was flourishing. Wheat crops were excellent and the mill's production was at its height. Wennery was a good miller, well liked and producing a good grade of flour for the folks around. It was a whole grain flour, unbolted, containing all the nutrients of the grain itself. It is said three large sacks of wheat when ground made one large sack of flour, and out of this the farmer paid the going rate of 1/8 to the miller as his fee.

Other communities around the area were developing. The town of Pomeroy which came into being in 1878 had its own gristmill which had been financed by Joseph Pomeroy, built by Wm. C. Potter,with him and Ben Day operating it. Pataha City, three miles up the Pataha Creek from Pomeroy, was an active community with a new gristmill, operated by water power and a turbine, with John Houser the new owner and operator. Luther Stimson was back up in the Alpowa country and word came that he had just completed building a gristmill at Alpowa and was now living there with his family and operating the mill as its miller.

In 1881, the Columbia Center gristmill changed ownership again. This time, A.B. Bean sold it and the accompanying land to Frank Paine of Walla Walla for $2000. This was in June 24, 1881. This made no difference in the output or operation of the mill for Wennery was still the miller. He and his wife had just had a little child, a brand new little girl and their happiness in the infant was shared by all their neighbors and friends.

Men coming to the mill with their wheat to be ground, often sat around and visited with Wennery or another farmer who was also in with his wheat. While the creamy-colored flour was being ground, sifting out through the cracks in the stones and filling the air with its pleasant wheaty smell, the farmers followed a custom established when the mill first started running. They would inscribe their names on the round, smooth center support post in the mill. This served as a record of their visit to the mill, and along with their names and dates was added the amount of grain they had brought to be ground Some scratched or carved their names into the

Graceful center post on which was posted a record of farmers' visit to mill, farmers signing their names and dates. Post was painted white for this purpose and the signatures covering it entirely are still faintly visible. Note straw around base of post. The mill is used as a barn and storage for hay.

wood with a penknife, perhaps with the miller's knife he used to cut the twine for the sacks of flour. Others wrote their signatures with a sharp pencil.

Today, in the 1970s, the 'shell' of the old mill still remains standing and the post is just as it was with the exception that now the signatures are dimmed by time and faded almost beyond recognition although a few still show. They are all written closely together on the smooth pine post with its coating of white lead and raw linseed oil.

Wennery was from the East and it was common practice there for customers of a mill to "post their names" wherever there was a smooth surface to record them. Notebooks, ready with neatly lined paper pages and a spiral wire binding to hold the pages together were not even dreamed of, so names would be written on the floor, or wall, or on a door, wherever handy, the customer taking care of this while the miller 'tended the stones'.

At the Columbia Center mill long ago, none of the farmers recording their names would have thought of themselves as making a decorative object, but that is what resulted for the post is an eye-catching, name-decorated attraction.

"Put it on the post", the miller might sing out as he weighed or measured the wheat and started emptying some of it into the hopper above the stones. He'd pull a lever and the big, well-balancedstones would begin turning as a little stream of wheat poured down from the hopper into the center of the top runner. The stones would grumble and rumble and begin shaking the big, sturdily braced room. The gears would make a loud grating racket and flour dust would seep out from around the stones and begin to fill the room, its appetizing smell reminding the farmer it was just about eating time. Soon the miller might reach down and take a little of the flour between his fingers and thumb, rubbing and testing it, like a cook "feeling" the salt for her soup, knowing when it was right by the touch.

It took about a half hour to grind a bushel of wheat and when it was ready, the cream-colored flour contained all the healthful parts of the wheat kernel. After the housewife took it home, she used it up fairly fast for its keeping qualities were limited with the oil of the wheat germ in the flour. But with a big family of seven or eight children, and usually a hired man besides the housewife's husband and herself, a sack of flour from the mill didn't last very long anyhow and she was rarely troubled with the thought that the flour would not keep for too long a time.

The signature, "Enoch Long, 1878", is still legible on the post. Beneath the name are other notations recording the amount of wheat brought in, but it is too blurred to read. "A.M. Haworth, April, 1879". "J.J. Miller", date blurred, "Clark Long, March 12, 1878". Haworth's grandson came back to the mill in 1957 and reidentified his grandfather's signature.

Other farmers in the region who would have brought grist to the mill and signed their names were Bankson, Cluster, Calloway, Greer, Buchanan, Baldwin, Hubbard, Hutchens, Hunt, Hoppe, McCormack, Day, Butler, Morse, Long, Hastings, Williams, Bosley, Slaybaugh, Mohler, Holbrook, Ashby, King, Morrison, Flerchinger, Herres, Dick, Brooks, Scoggin, Sommerville, Jennings, Carpenter, Mitchel, Ward, Gilbanks, Deardoff, Teale, Harris, Armstrong, Leathers, Oliver, Light, Chester, Hubbard, Gill, White, Bean, Rush, Surrat, Travis, King, Gallagher, Rose, Rathbun, Spallinger, Davis, Russel, Hunt, Plummer, Jennings, Warren, Able, Graham, Shawley, Downing, Buchanan, Carpenter, Humphrey, Courtney.

The gristmill may have enjoyed other experiences occasionally. Perhaps on warm May nights, young couples hand in hand strolled down to it, threading their way among the firs and stepping over the big wooden flume, the young man helping a pretty girl. They visited the swampy area where the old water wheel stood silent, resting for the night, and the young pair paused in their embrace to watch in wonder as the moonlight caught the water still dripping from the wheel's cups, turning the drops into sparkling diamonds. Many a lover's secrets, the old mill knew, and many words spoken in soft caressment the old mill still holds locked safely in the depths of its wooden recesses.

About the year 1885, a change occured in the community and in the Columbia Center gristmill. Mr. Wennery had been handling the milling operations with fine success and he and his wife were a happy family, well-liked in the community. Their little girl was a delightful child, with golden hair and bright blue eyes.

But tragedy crippled the region in the form of a contagious epidemic of diptheria. There was no stemming the spread of this dread disease and before people quite realized its nature enough to keep their youngsters away from others possibly carrying it, many children were afflicted. Doctors did all they could, still several children sickened and died from the disease.

The Slaybaugh family lost three children. The Moffatt Williams lost one child and the Hutchens family lost their second boy. The little Wennery child was stricken and died, and many other children were lost to families helpless to save them.

From then on, the light seemed to go out for the Wennery couple and they felt they could no longer bear to live out here any more, longing for the closeness and tenderness from relatives "back home". Wennery gave up his milling job at the Columbia Center mill and he and his wife returned to Missouri to live.

After this, the grist mill having lost its miller with no one in steady charge to keep it up and take proper care of the machinery, as well as grind farmers' wheat into flour, it gradually went into disuse. By this time, and perhaps before, the mill had been steadily decreasing in its output and business, and this may also have influenced the Wennery's in their decision to leave. Houser's mill at Pataha City was producing a finer, whiter flour, and many of the farmers' wives preferred this to the Columbia Center flour which wasn't bolted as Houser's was to remove the middlings, or bran and shorts. The white flour made a fluffier loaf of bread, and this was an attraction which brought farmers to Houser's mill with their wheat. Although Houser charged them a little more than the 1/8 always paid at Columbia Center, the farmers were willing to pay the 1/3 at Pataha City for Houser gave them the bran and shorts which they could use for stock feed and which just about made up for the difference in cost.

The Columbia Center mill stood idle for many years. Moreover, the little town itself by the middle 1880s had begun to decline. Stores were closing down, and the population of the town was dwindling drastically. The railroads had come into the country in 1886 but Columbia Center was bypassed, suffering as a result from lack of business and attention.

In 1902, the mill property changed ownership again. On December 30, 1902, N.O. Baldwin bought the property, which included the gristmill lots 15-16, block 2 from Frank W. Paine of Walla Walla.

Oscar Baldwin was then sheriff of Pomeroy and a real estate dealer. When Mr. Baldwin purchased the property he probably bought it outright, but on October 14, 1911, he decided to place a mortgage on the property with the Northwestern and Pacific Hypotheek Bank.

Vanished Gristmills

Columbia Center gristmill in its original location beside the Pataha Creek, but now no longer used as a flour mill.
(Courtesy Robert Beale)

It is thought that Oscar Baldwin did not reside on the property at Columbia Center himself but that his son, Coil Baldwin lived there. Coil raised cattle and used the old gristmill as a barn, the cattle gathering in the lower stone part of the mill for protection or at the areas around it to graze or drink from the water in the creek running beside the old mill.

Columbia Center gristmill
(Courtesy Mrs. Wyoma Duckworth)

Gradually, the inner parts of the mill either rotted or broke from rust or disuse. The millstones were removed and one was placed for a time in a spot at the courthouse, with the other said to have been crushed. During the World War II days when scrap iron was needed, any metal shafts or gears in the mill were taken out and given to the war effort. The old water wheel which had tongued up the creek's water for so many years, quickly decayed, the moisture aiding in its decomposition. The mill ditch though not used remained visible on the hillside above the Peola road, and the dam area no longer performing a necessary function, gradually reverted back to brush and trees. The boards and other lumber from the box flume were salvaged for farm buildings.

In the early 1900s, the little town was scarcely a remnant of what was there before. The old gristmill, still sturdy and strong, had become useful as a part of the farm, serving as a barn. The Baldwins would not have considered tearing it down. There was something endearing and nostalgic about the big, old cumbersome structure. They liked to have it there and when anyone visited the family at the farm, the visit wasn't complete until the guests were shown the inside of the old mill building and had it all explained how it was, long ago.

These were hard times for farmers. Banks were foreclosing one farm after another; too hastily, undoubtedly. Any farmer needed every leniency he could have to be able to meet his mortgage payments. No bank in a farming community should

Vanished Gristmills

Columbia Center gristmill about 1931. At this time mill was being used as a barn by the Frank Spalingers who were renting the property from the McCown brothers. The buildings in the background consist of a residence and its buildings. The gristmill stands in its original setting here beside creek. Box flume came down incline back of the building where trees show. **(Picture courtesy Anna Baldwin Butts)**

This is the schoolhouse taken in 1931 when Anna Baldwin Butts was the teacher. This location is across road from gristmill. Jackson's sheep were being driven past woodshed and barn. The schoolhouse 'privy' is beside the shed. **(Courtesy Anna Baldwin Butts)**

have been hard-nosed in collecting payments which were so often hard to come by, ruthlessly taking away lands when a payment could not be met. Such happened to the farm property on which the old Columbia Center gristmill stood.

On January 25, 1915, the bank foreclosed, taking the gristmill site, lots 15-16, block 2, and other property.

The old gristmill was now on precarious footing, so far as its future was concerned. Thus far it had withstood time under kind owners, but everything on a farm must earn its keep, even the shell of an old gristmill, and the remains of this old mill, although still as stout as ever, was not entirely a success as a barn. Its safety could have been in jeopardy, for new owners might dispose of it as soon as they took it over.

Fortunately, the old gristmill's existence was assured, with future owners not only appreciating its worth but also making it a useful part of the farm. On November 13, 1916, the old gristmill was again in secure hands when Oscar Baldwin, with his son Wyrn Baldwin's name on the record books, redeemed the property "which included lots 15-16, block 2, the gristmill property and rights to the mill, for $34,966.51."

From then on, property changes which included the gristmill moved fast and often, with the mill remaining intact through all the transactions. On December 5, 1916, the property and gristmill, in Wyrn's name, was sold by Oscar Baldwin, to Frank McCown and Emory and Malcolm, brothers.

The property operated under the McCown brothers, the gristmill shell still a barn for hay and cattle, until December 8, 1930. At that time a quit-claim deed went to Emory and almost immediately Emory McCown sold the mill and other property to J.W. Taylor.

The following year, November 6, 1931, J.W. Taylor sold the property to W.M. Taylor, a member of his family.

In 1932, W.M. Taylor sold the property to Henry R. Dixon and on March 23, 1943, Henry Dixon quit-claimed the property to Levi Dixon, a bachelor, and on April 10, 1944, Levi Dixon sold the property to L.G. Duckworth.

The Duckworths lived on the farm and operated it, also raising cattle and using the old mill as the others had. After Mr. Duckworth's death, Mrs. Duckworth, in 1975, sold the property to her son, Garry Duckworth. Garry and his wife and family, the present owners, live in a charming new cottage built "where the schoolhouse was", raise cattle and use the old mill as a barn.

The Duckworths in 1945, made a change for the better in the structural welfare of the old gristmill. It, in its original setting beside the creek was beginning to settle and lean badly from a soggy underfooting, becoming unsafe and almost unuseable.

They took the upper wooden portion of the structure from its rock foundation and moved it to a new, higher location on the opposite side of the creek. Here the mill remains intact and safe from the water, functioning as a dry storage place for hay and stock feed and "earning its keep" as a working unit on a busy farm. Additions have been built around the mill for stock use, and stoutly constructed metal fencing for stock surrounds the building. The old mill, center of all this, rises above its wooden skirts and like the old patriarch it is, stands proudly as if surveying and directing all barnyard operations.

The remains of this old gristmill, soon one hundred or more years old, is a fitting ornament for this setting near the Blue Mountains. Luxuriant fir trees still surround the little valley, and the cool breath of the mountain sweeps gently through the area carrying a nostalgic memory of long ago when Columbia Center was a lively, bustling town and the old gristmill was a significant and popular business, grinding the farmers' wheat into flour.

In 1945, the Duckworths moved the shell of the gristmill to the opposite side of Pataha Creek. They built additions around it and used it for a hay and stock feed storage area for cattle. The original siding is still on the building and the interior is now as sturdy as it was one hundred or more years ago.

Garfield County Gristmills

POMEROY FLOUR MILLS

In the middle 1870s, the stage coach line from Walla Walla to Lewiston stopped about thirty miles enroute at the Joseph Pomeroy stage station on Pataha Creek. As the people riding the stage coaches neared the Pomeroy stop, they felt a tingle of anticipation for they knew a delicious dinner awaited their arrival. Dinner was served at twelve o'clock noon, just as the stage arrived, and it was really *dinner* in those days; not the *1970s'* version of a ham sandwich and a glass of milk, or a piece of apple pie and a cup of coffee, but a full-sized, well-cooked, old-time square meal. In those early days, a ride of thirty miles took several hours, even with a spirited four-span team whisking the crowded coach of seven or eight passengers over the country side. Roads were bumpy, narrow dirt thoroughfares, dusty in summer and full of muddy, or frozen chuck-holes in colder months. Several hours of traveling like this could produce a roaring appetite.

A little settlement had formed near Pomeroy's stage stop. The Owsleys were five miles west, Parson Quinn and the King brothers and two or three families lived fairly close. Farms were developing on the higher places.

For some months Joseph Pomeroy had been mentally struggling with a few vital facts. There was the matter of the long distance to Dayton which was an inconvenience to people living near the Snake River. Columbia County comprised all the area then and folks up in the Pomeroy country wanted the county seat relocated closer to them.

About eight miles distant from the Pomeroy stage stop, and near the Blue Mountains, a little settlement calling itself *Columbia Center* had valiantly organized itself and become platted into a new village. Pomeroy was sure they were courting the hope of becoming the new county seat.

Joseph Pomeroy subsequently made up his mind to plat the property around his stage station into a town. He hired a surveyor for this in 1877, and suddenly, the little settlement around the Pomeroy station became the new town of *Pomeroy*. Joseph gave away lots to anyone who would start up a business for he wanted the town to develop fast. He was sure it was the best location for the new county seat, with a likely chance for selection.

Joseph was pleased when a William C. Potter, anxious to build a gristmill, was drawn into the new town promoting such an idea to all the farmers and new citizens.

Vanished Gristmills

It wasn't difficult to enthuse a farmer about a gristmill being put up near his farm, for this was a convenience as well as a necessity he and his farmer neighbors all wanted, and needed.

Fortunately, a gristmill could be easily built at any location where a stream had a good fall from a higher to a lower level, with the power from the stream a cheap means of energy. In nearly every little hamlet in the eastern part of Washington, one gristmill after another was setting its paddles to lapping up the waters of a town's creek. These gristmills were often being built less than fifteen miles apart, but fifteen miles traveled by horse and wagon was a long ride in the 1870s and 1880s. Farmers did not have the time to spend driving a heavy wagon load of wheat to a gristmill several miles away in another town to have their wheat ground into flour.

Potter took on the building of the mill, but since he had no money himself to put into the project, Joseph Pomeroy contributed $2,000 and Mr. Heaton, a farmer in the area, is said to have put up the rest of the cash. Pomeroy also furnished the lots for the site of the mill in the western part of town and assigned the necessary water rights from the creek. Before long, the gristmill was taking shape along with a number of other business establishments soon ready to be stocked with goods and opened up.

When the gristmill was finished, Ben Day, a leading new merchant in the fast developing town, operated the gristmill as its miller. According to an 1878 issue of the Dayton *Columbia Chronicle,* stating, "The mill is a very successful one, operating day and night." The gristmill was not lacking in business from the farmers.

The gristmill was a two-story, wooden structure equipped with one run of granite millstone, and was operated with a turbine which took water from Pataha Creek.

The water to the mill came by way of a long, wooden, box-flume coming several hundred yards away from the Pataha Creek. The wooden millflume lay along the side of the hill—Sunnyside Hill—like a long, grey snake sunning itself, the flume's weathered boards blending in with the grassy slope. It rested on squat, wooden supports just high enough for the flume to clear the ground and have an adequate tilt. The flume started from a mill ditch which took water out of the creek. The ditch crossed what is now the main highway through the town, at a location occupied by a trailer business today. The water, after running its long course in the box flume on the hillside, formed a dam at a point above the millsite "near the Jeffry Williams place". From the dam, the water flowed down a millrace beside the St. George (Dr. Sherfey) home, and into a standpipe to the turbine.

The millpond or dam that formed at the end of the wooden flume on the hill became a swimming pool for anyone who wanted to use it during the hot, summer months, and before a community swimming pool was put in at the city park. The millpond was about 30 X 50 feet in dimension, and it is said a large willow tree was right in the middle of it. Many people nostalgically remember that they enjoyed many hours of fun swimming in the pond.

There is an amusing detail connected with the old mill ditch and the water running through it. A Mr. Hender who lived near by, obviously intrigued with water and water wheels, like any small boy playing in the springtime freshets with tiny water wheels, conceived a novel idea for watering his garden. He obtained an ordinary wagon wheel with the shaft, or axle, still in it, and put this wheel into the water.

At spaced distances around the wheel he attached ordinary cans which had once had canned goods in them. As the wheel turned, the cans filled with water and when they reached the top of the wheel and turned upside down, they emptied themselves into a trough which he made just for the purpose of catching the water from the cans. This made a little stream which he directed along the trough to his garden where the water ran in rivulets down every row. It is said he had the best carrots, lettuce, onions, cabbages as well as roses and phlox of any garden in the country.

In the 1890s, Mr. Heaton who is said to have owned the mill then, sold it to Mox Houser. Mox was one of the sons of John Houser who owned and ran the Pataha City mill three miles up the Pataha Creek. Mox put roller equipment into the mill and a simple bolting apparatus to sift the flour. It is thought that the millstones were not destroyed, but that someone from Dayton took them. Whether this was as a private collector or for use in one of the Touchet Valley mills, is not known.

In the year 1903, the original Pomeroy mill burned down. Carrol Grafton Shawen, father of Ralph Shawen of Spokane, was employed at the Pomeroy Mill when the fire occurred, although he was not in the mill at the time. Graf (as Carrol was usually called) had received minor injuries from having had a sack of wheat fall on him and then more or less "burying" him under wheat when other sacks followed and pinned him underneath them.

After the mill burned down, Mox Houser rebuilt it into a large, three-story structure, equipping the mill with roller machinery and all the best cleaning and bolting equipment available then. He installed a new turbine which still would use the water from the creek as before. The old turbine was retained and a short time later, these turbines were used to produce electricity in the mill for a number of years.

Pomeroy Flour Mills, Pomeroy, Washington, owned by Mox Houser

When Mr. Houser built the new mill he changed its location from the original site which was located closer to town. Mox is said to have bought two or three houses which were on Main Street, moving them farther east and across the side street to make room for the new mill. The new mill was on Main, across the street and a little west of the Dr. Sherfey home, with the houses directly across the street. One of them was a two-apartment building which Mox called "The Ugly Duckling", the name painted on a sign above the door.

A miller by the name of Rollie Kribbs ran the Mox Houser mill for several years. The flour was called "Pomeroy's Blue Mountain Brand", and the flour sack had a picture of the Blue Mountains on it.

The flour from Mox Houser's *Pomeroy Flour Mills* was in some competition with that of the elder John Houser's *Houser and Son Flour Mill,* (another son, William J. Houser, in partnership with his father). One farmer said he was swayed to the use of John Houser's flour because it took a prize at a fair when it was exhibited there. Both products, however, were of equal high quality.

Garfield County Gristmills

During the years when Mox operated the flour mill, he became greatly interested and involved in wheat speculation. At the time of World War I, this enterprise became so extensive he rented a room in the bank building in Pomeroy and carried on these operations there. From this business, he became wealthy.

At one time he owned the Portland Hotel in Portland, Oregon, as well as a number of other large holdings. Just before the outbreak of World War I, he had the business foresight to buy up all the leasing contracts of Pacific Coast transport boats, which gave him control of all shipping. The government, in order to use these ships for transport abroad had to lease them through Houser for they were under his control. This business added to his wealth which is said to have been over three million at one time.

Mox Houser during this time still continued his wheat speculation, buying and selling, but this venture suddenly took a downward spiral at the end of the War. As so sadly often happens, his wealth also plummeted and he lost everything. All his business and other properties went into bankruptcy and were sold at auction.

The Cardwell family in Pomeroy became owners of the Pomeroy Flour Mills in the 1920s, and when the Cardwell estate was later settled, "Frank (Cardwell) took the mill, and Hugh the store", according to the late Mrs. Cardwell.

In later years, the mill no longer making flour, the property became the *Pomeroy Warehouse and Feed Company* under Leonard Herres. The buildings of this company were destroyed by fire July 23, 1950, the blaze starting early in the morning. Mr. Herres rebuilt, putting up a three-story chop mill, and a large sack and feed warehouse, constructing all these buildings of fireproof metal.

The main motive for the town of Pomeroy's early beginning was soon lost in its endeavors for its own purposes, the working for the benefit and needs of the farmers and people living there. In 1881 and 1882, this little town weathered all conflicts to win the county seat controversy of the newly established Garfield County, and has remained the county seat ever since.

Wheat has always been this region's greatest product, and at one time the production of flour ran a close second, for Garfield County's flouring mills ranked among the best in Eastern Washington.

The Houser men who ran these flour mills have often been noted as "men of genius", and they unquestionably belong in the annals of great men of industry for that period.

Vanished Gristmills

GARFIELD COUNTY GRISTMILLS

HOUSER FLOUR MILL, *Houser and Son Flour Mill*
Pataha Flour Mills

Garfield County's five early grist mills serving scattered farm families in the communities in the 1800s, were an important representation then of the flour milling industry in Eastern Washington. None of these mills sought to compete with each other for they were built entirely from necessity for the needs of the people in each mill's locality. Four of these mills knew only a few years of activity, but the fifth, the *Pataha Flour Mills* was to run many years and surpass all others throughout the region in productivity and excellence.

In 1877, Garfield County (then still Washington Territory), suddenly leaped into an industrial awakening. Sawmills dotted the banks of the Pataha Creek almost from the moment it sprang out from the cool edges of the Blue Mountains. In 1877 and 1878 two grist mills were begun, and also the little towns around them. One was in the mountain valley of luxuriant fir trees at Columbia Center, and the other in the treeless, deep-soiled valley at Pataha City along the Pataha Creek. Pataha City, bordered with bunch grass hills and rimmed at the top with rich wheat fields, was a natural setting for a town and for a gristmill.

In 1878, John Bauman and George Snyder, two skilled millwrights, began the construction of a large gristmill at Pataha City, and the mill soon came under the management of John Houser. He took control of the mill as part owner with other partners in 1879 and by 1880 became its sole owner.*

John Houser had received expert training in milling as a youth in Germany where his brother was in the milling business. John learned this trade in the time-honored custom of going from mill to mill throughout the country and working in each to gain wide knowledge in the field. By 1879, John was well experienced in this trade for since the year 1866 he had been in charge of milling at the Meyers Falls Mill at Colville, distinguishing himself there by developing and producing several creditable "firsts" in the milling industry.

But before going either to Colville, or Pataha City, John experienced a life that was full of adventure, which he neither planned not anticipated when he came to this country from Germany in 1848. These adventures could only have happened in those early days of Indian uprisings in America and during the first feverish scrambles for gold in 1849-50. John was swept up in all this but he came through

*One of the partners was Mr. Harford, who helped finance mill at Pataha.

unscathed and even achieving success in the gold fields. Yet his main purpose always was an unbending desire to establish himself in the pattern of life he had set out for himself long before the time he landed as a youth in this country. This entry into the United States was a notable one.

When the large freighter docked in the New York harbor, a raw-boned, lean young man of nineteen waited his turn, one of the last, to walk down the gangplank, his gaze riveted on buildings before him rising higher than any he'd every envisioned. Gripping a shabby, linen valise, he took a worn slip of paper from his inner pocket and glanced at the words, although he already knew them by heart: "Dear Elsa—This is John Houser, Please give him a bed and...." John stuffed the paper back into his pocket. At least, there would be a place to go and he would be able to find it from the explicit directions given him. Then he would begin looking for work. Although he scarcely knew a word of English, he felt he would be able to find employment without difficulty. He was well prepared in a time-honored, important industry, that of flour milling, and he knew there were many gristmills in the eastern part of the United States.

John found it good to step out on land again. He had been on the boat a long time, but he'd do it all again if he had to. Coming to America had been his dream since he was a small boy and when the opportunity to work for passage on a boat finally opened up for him, he grabbed it quickly. It had been 300 days of hard work, and resting only a few hours at night. Usually the labor was on the cold, sea-sprayed deck, lifting and handling cargo, or beneath the holds, doing equally demanding work. He wasn't as big a man as some of the others working, but he could match any of them lifting and shifting the mammoth loads. He never told anyone about his left arm although the pain from it often nearly blinded him, it hurt so much at times. He made it perform just as hard as his right arm, doing as much work as anyone. Maybe some day he'd recover its full use but now it was almost as bad as when it became badly crushed in his brother's mill a few years ago.

The first thing now was to find work. John started out almost immediately looking for employment in a gristmill. The first one turned him down, as did the second. Here the mill owner's eyes coldly appraised the rough-looking, shabbily clothed young man who spoke unintelligible words accompanied with wildly gesticulating hands, for John knew only a little English. When John was first turned down, he thought it only because work was scarce. Time after time, with the same thing occuring he knew he was at fault in some way for his efforts to please and impress a would-be employer bought only rebuffs, and his attempts at English sometimes ridicule and laughter.

Pataha Flour Mills. This mill still intact although not running since 1940. Equipment is still in the mill. **(Picture courtesy Robert Beale)**

True, his appearance was against him. He had begun the journey from Germany almost a year ago clad in a pair of dark, cotton, home-spun trousers, wearing one of the two faded shirts he'd brought, and a badly fitting jacket once owned by his brother. This wardrobe did not improve in appearance after being exposed to wind and weather and sea spray for nearly twelve months on a boat, wearing them as he worked for his passage. His sparse, medium-sized figure did not give the assurance of physical strength even though his blue-grey eyes burned with energy and ambition. His seeming uncouthness and illiteracy from lack of ability to communicate, worked against him, hiding his real worth. He found little, if any employment in the field in which he not only would have excelled giving more than his measure, but possibly might have brought excellence and acclaim to his employer's name as well.

Working at any odd jobs he could find in New York to support himself, he also searched for a sister whom the family had lost track of. Tragically enough, much later when he was no longer in the eastern part of the country she died, leaving her property to him. By then, he could not be located and the property finally reverted to the state of New York.

He occasionally found a little work here and there, at one time securing employment on the river boats of the Mississippi River. He heard about an immigrant outfit bound for the California gold fields, and hastened to that town to join the outfit. The rough, good-natured men laughed heartily at his German-flavored English, but he found there were other foreigners there too, and the broken English spoken among them interwoven with Swedish, German and Italian soon became common-place talk, with John understanding them as well as learning English quickly. The captain of the band recognizing a natural authority in John's makeup, gave him a team of oxen to drive, telling him to say "Gee" to go right and "Haw" for left.

Word had come of hostile Indians when the outfit was in Salt Lake, but thirty-three members of the outfit John was with, felt they could manage and decided to go on anyhow, with Houser one of them. These men arranged to separate and then meet later at the Humboldt River. They met there as planned, but they also met a howling band of four hundred Indians, including women and children, who began attacking them from all sides with every weapon they could lay their hands on. The attack was swift and fierce, but suddenly they all rushed away as quickly as they had appeared.

The party of men still were determined to push on ahead although the Indian battle was nothing to be taken lightly, for they would undoubtedly attack again. Exhausted, Houser and his group stopped at the river for a time to rest and eat, but the captain urged them to continue on, eat on the way and make a rush as fast as they could for Stony Point, a place several miles distant where he knew they'd be safe. The tired men were reluctant to move on, lingering at the river. That night Houser was placed on guard. He listened and watched intently for hours, hearing nothing unusual. Around midnight, he was sure he could make out in the darkness, threatening shadows moving about, and could hear unrealistic bird and animal cries. He quickly alerted the others and they flew to the horses hastily harnessing and hitching them up, and starting off at a wild gallop for the point.

In Houser's words, "We got nearly to the point when the Indians caught up to us. They had about a dozen muskets but they shot arrows at our horses. My feet got heavy and my legs wobbly. The Indians were making a frightful noise. They were all coming after us. My feet didn't carry me any more and I had to hold on to the wagon. I never was in a fight before and couldn't hold a gun or revolver, but after the fighting got started I found I could stand up and do my best like the rest. Some of our horses were killed and we had to hitch up others. While we were hitching the Indians began to roll big rocks down on us. We kept on moving and fought for about half an hour. How many Indians were killed I don't know; it was not tens but hundreds, women and children too, and all were scattered around. At last they broke and ran. We had ten or fifteen wounded.

"But the danger was still there for as we reached the Point, the Indians had set a high brush fire blocking our way. The fire was burning on both sides of the road, just like an arch (of fire), the flames coming together on top. I could see some of the men through the arch. I ran through and my clothes and hat were burning. The captain told me to roll in the sand and that put it out but my face and hands were blistered. When the fire died out a little we found thirteen of us left out of thirty-three. We had no water and couldn't get to the river. Our tongues were so swollen up we couldn't move them. We traveled about a quarter of a mile and came to a place where there had been a spring but with nothing there now. We hadn't a single thing to eat and were about 150 miles from any place. The captain said all we could do was to travel and if the Indians caught us they would kill us for we had no more ammunition.

"The next day we found some dried fish but it was too rotten to eat. We ate roots of different kinds of grasses...We went down to the river and drank a little alkali water. We got so we couldn't sleep but we kept going. As I went along I got as light as a feather. The hunger kind of went away. It was that way three or four days and on the fifth night I saw a light. It was a trading place and when I got there I found three or four fellows in a cabin, but the trader there made us stay outside and rest, and finally he gave us just a little food. We were frantic from starvation and I once got up and ran against the door pounding on it and begging for food. The trader finally took us into the kitchen where he had made a big kettle of soup, and he gave each of us a small cupful. We begged for more and then he explained to us that if he gave us anything more to eat we would all be dead in two hours for our systems were so starved.

"In a short time the rest of the party staggered to this door and the trader gave them a small amount of soup just as he had given us. In the morning the trader sent out searchers to find the rest of the party and finally brought in the last man who was so weak he had to be held on a horse.

"We stayed there ten days and after that we were in shape to eat. We had all the beef and soup we wanted. Beeves were cheap and we cooked and ate all day long for over a week."

After John reached California he became involved in many mining ventures, sometimes winning and often losing. He went to Williams Creek in British Columbia and there in the early 1860s, involved in gold mining, acquired $80,000. Soon after this in another venture which went "sour", he lost most of this because "he was too honest to duck or protect himself by subterfuges."

A poignant incident touched him in a personal way. He knew that his family back in Germany would never be able to understand the course his life had taken, and this was brought home to him in a touching way. He sent his mother a draft for $1,000 with a brief letter of love and remembrance, "Mama, buy yourself something nice, and don't worry, I can afford this...." In a few months he was surprised to find a letter in his box from his mother, for she never wrote. In the envelope was his $1,000 draft, returned with the admonition, "You have not been in America long enough to earn that much money honestly." He had forgotten that five dollars was an average wage for a month's work in Germany, and that to his mother, only a rich man, or a scoundrel could have $1,000.

John was never able to recoup his fortune in Canada although he returned there for a third time to try. He went to Seattle and worked at Whidby Island, still hunting for gold. This was at the time when Seward purchased Alaska. The army called for men to accompany the troops going in to take possession and Houser was one included in this group. He and his companions whipsawed lumber for army barracks for a period of about two months, "after which about a half dozen of the men went to the far-off Frazier River gold fields."

Here for a while they made money as fast as they wanted simply by picking up nuggets off the river bed. There were many adventures connected with his short stay in Alaska but he did come out with some money (about $3,500), for the government allowed the men to keep all the gold they found there.

The nugget-collecting had to stop when hard winter set in and the river became frozen. But one man had an ingenious method for solving this set-back in their operations by getting them all working at cutting down through the ice, removing it piece by piece to get at the pay-mud below. They could not wash this soil now in the winter but would save it for later, shoveling the mud into buckets or any other container available and carrying the heavy loads up to their shelter, a primitive cave-like abode. Here they dumped the mud on the hillside beside their cave, believing it would be in a safe place until the coming of spring, whereupon they intended to wash all the soil for the nuggets, undoubtedly there.

If Alaska had a gentle side to its nature, it did not show in her winters, nor even in her emergence from its cold embrace, for at the first indication of spring she roughly thundered into the region like a wild giantess. She broke up the ice and let the pent-up water under it wreck its vengence on anything it chose. The river's first objective was the huge pile of mud, her very own possession. She flung the big blocks of ice at the men's camp with a mighty roar like a first howl of life, and sent torrents of waves up along the little mountain-side of pay dirt, washing it swiftly back to the river's bosom. The men, white-faced, watched trembling in shocked silence glad only that they had been spared.

Houser still had his $3500 and he knew he would not lose it, his ambition to become established in milling now coming again to the forefront, bringing him back to the states, through, at least for the time being with Alaska mining ventures.

In Durham's *History of Spokane County and Inland Empire,* a letter from W.P. Winans (early resident and historian of the Colville region), states: "That fall (1866) a number of men came down the Columbia from the French Creek mines, among them John Houser, a miller, who was employed by Mr. Meyers in December, 1866."

The Colville mill on the falls of the Colville River, which became known as the *Meyers Falls Mill,* was the old Ft. Colville mill and significant historically as Colville itself was historic as an early settlement in Eastern Washington, figuring so widely in the fur trading, trapping and exploration days. The Meyers Falls Mill was acquired by L.W. Meyers in 1866, he and George B. Wannicott leasing the property.*

Later, John Houser in a letter to Mr. Winans, reminisced: "I came to Colville in the fall of 1866 and went to work for Mr. Meyers and we remodeled the old Hudson's Bay Mill". Winans worked also for Mr. Meyers in 1867 as bookkeeper in the mill, and recalls in his memoirs that John Houser with the help of Lewis Meyers, who was

*Details on this in *Stevens County Gristmils, Colville,* this book.

a carpenter and cabinet maker by trade, rebuilt the mill with Houser constructing a "middlings purifier", the first one ever built or used in Eastern Washington. This was a system of combining bolting cloth, preferably of silk, with air currents to sift the flour and obtain a finer product. The bran, the outer husk, and the middlings, the inner, coarser parts of the ground wheat were collected separately. Flour produced in the manner was called *Patent Flour*.

Houser was spurred along in his desire to produce a fine grade of flour by M.M. Cowley*, a trader at Cowley's bridge and later banker in Spokane, who, visiting the Colville mill joshed Houser about that "dark color of your flour". "Too bad," he quipped, "You can't make as good flour as they do in Walla Walla."

"That's all right," John retorted. He stood a moment looking at Cowley, lifting his left arm and placing his forefinger against the side of his nose, as if such a gesture might help him think a little better. "That's all right, but you just wait 'til I get my machine built, and my bolting silk here. You'll see whiter flour than you could ever get any where."

"That so? Well—now, I'll just bet you a new suit of clothes it won't be as good."

John Houser got his suit of clothes which came from England and 'around the horn', for after the bolting device was finished it did give flour of an excellent white quality and fineness.

Houser instigated another first at the Colville mill, the initial placement on the market of a cereal known as Farina. John's letter to Winans further recalled: "We made the first patent flour in 1867, and the first farina in the United States." When Houser exhibited his patent flour and his farina at the Oregon State Fair in Salem in 1868, both products won first prize.

The following summer after being employed at the Meyers Falls Mill, Houser went back to the mines, but he returned to Colville again that winter. He then went to work for Samuel and Joseph Oppenheimer who had just acquired the Pend Oreille mill. The Oppenheimers had heard about Houser's work at the Meyers mill and employed him to remodel their mill, which Houser did, putting in a middlings purifier there. He then manufactured farina and a patent flour for them for a time.

The farina which Houser developed at the Colville mills became so popular, that wagon loads of it were taken to Walla Walla merchants, shipped to the firm of Alden and Lewis in Portland and sold in large quantities to Ft. Sherman at Coeur

*M.M. Cowley, no relation to Henry T. Cowley, early pioneer of Spokane, teacher and missionary.

d'Alene. It also found a ready market among the housewives and also the hordes of men laying in supplies on their way to the Idaho mines.

Although John Houser had made some money in his mining ventures, he yet had not found a suitable investment for it since coming to Colville. The only business venture he would consider was a flour, or gristmill of his own, and he was constantly on the lookout for one.

He was a single man with no family here. He was particularly fond of a young woman, Miss Marie Reiniger, whom he had known in Germany and wrote to often, knowing she would play an important role in his future life. He also had a friend, Jacob Bilhmaier, from Germany who had come to the United States before him. They were near the same age, John born May 15, 1829, and Jacob, 1836, and both from Wurtenberg, Germany. John had almost lost touch with Jacob during his gold mining days, but learning that his friend was fairly close by in southeastern Washington at Pataha City, he wrote to him. Jacob had secured property in Pataha City on which was a strong spring and an excellent creek, both of which took Bilhmaier's name, and this Bilhmaier creek was to play an important role later in both these men's industrial endeavors.

John stayed with the Meyers Falls Mill until 1872. At this time the old mill was showing signs of decay in vital sections of the foundation and other places too serious to ignore or even try to repair. The timbers inside the building and most of the planks and boards were still in good condition. All the mortised joints held as sturdy and solid as when first fitted together. Meyers decided to close down the mill in 1872 with plans to demolish it and build a new one. This he did, using some of the flooring and ten-by ten timbers from the old Goudy mill. At this time, John still employed at the mill, decided to go back to Germany on vacation to visit friends and family. In a year or so he returned, bringing a wife, his former sweetheart with whom he had corresponded throughout the years.

After visiting in Germany John returned to Colville running the new Meyers Falls Mill, the location still on the beautiful falls of the Colville River. He continued working at Colville until the middle 1870s.

By 1877 or before, John knew through Mr. Bilhmaier that two skilled millwrights, John Bauman and George Snyder, were building a flour mill at Pataha City and that it could be bought from them. Before John made his decision to purchase this mill, he went to Spokane to look over milling possibilities there. He met and talked with James Glover about the industry, and Mr. Glover, anxious to have a mill started in his little settlement of Spokane, offered Houser the power rights to the falls for $600 if he would build a mill there. The Spokane country was not a wheat producing area then, and John, gazing out over the pine covered landscape without even a garden in sight, much less a wheat field, stood in contemplation a moment. Then pressing his left forefinger characteristically against the side of his nose, and peering intently into Glover's mild blue eyes, he said, "What good is a falls if there are no wheat fields." He then made his decision to go into milling at Pataha City.

The Pataha mill, a good-sized one, was finished in 1879, and John moved there to take charge of it. It is said he started its operation with one or two other partners but soon, 1880 or 1881, had it under his sole ownership, running it at peak operation under the name of *Houser Flour Mills.* It was a well built gristmill, equipped with two run of granite millstones, each of the four stones 36 inches in diameter. They were imported from France, which in those days produced the best millstones. There was a bolting apparatus, (some of these frames with scraps of silk mesh still clinging to them are in the attic of the old mill, which is still standing.) The mill had conveyor belts with cups to carry flour up from beneath the stones to bins, and a wooden auger to bring the wheat from storage areas to the cleaning apparatus.

John built a little bungalow close to the mill, surrounding it with a white picket fence, and he and his wife and family lived there. From the first, the flour mill was an industry of which the people in the little town and surrounding countryside were intensely proud. Farmers from miles around, often from very long distances would bring their grist to Houser's mill, even from other communities which had their own gristmills, "Just to have a sack of Houser's fine, white flour for baking something special." The Houser process removed all the husk in its bolting, whereas the flour ground between the stones without being sifted, contained all parts of the wheat. This was a coarser and darker colored flour and made a heavier bread, and, although some may not then have considered such a possibility, it undoubtedly also produced an extremely healthful bread.

Anyone bringing grain into Houser's had to pay a little more to have his flour ground there, 1/3 instead of the going rate of 1/8, but they considered it to be worth the extra price, for Houser gave them the bran and shorts and this they could use for stock feed.

The Bilhmaier Creek supplied the water to run Houser's mill by way of a turbine installed in the basement of the building. This machine powered all the machinery at the time the mill was first built, and although the mill was enlarged and added on to in later years, so far as is known the turbine was not replaced but remained the same.

A short shaft from the turbine ended with a wooden pulley-wheel about one foot in width. This was belted to a huge wooden power wheel at right angles to the turbine and about ten feet from it. The power wheel was strongly constructed of laminated, light-colored hardwood and was about six feet in diameter, with curved

Picture of Pataha Flouring Mill taken July 10, 1897. Men, l to r—Joe Waldher, Chas. Ward, John Gustafson, Henry O'Kinga. Women, l to r—Artilla Houser, Bertha Houser, Anna Waldher
Courtesy Garfield Co. Hist. Society

Front view of Pataha Flour Mill as it looks today. Original mill was to the left of the line where new siding was added for remodeled section, at right.

spokes mortised and fitted skillfully in place. A long wooden shaft extended from the power wheel's axle on which were a number of roller pulley-wheels of different widths which were belted to similar wheels overhead. These belted wheels activated other shafts and wheels which in turn ran all the equipment and machinery throughout the flour mill.

Later, when the mill was enlarged in 1891, the wooden shaft was replaced with a long metal one with additional pulley wheels to power the new equipment.

A dam on the creek about a mile from the gristmill supplied the water to run the mill. This water was carried along the hillside in a ditch which brought the water to a spot above the mill. There it poured downwards into a wooden standpipe. The force of the water was often so strong, high spurts of it shot up fountain-like through holes and cracks in the wooden pipe. The standpipe connected to a 16 inch pipe leading to the turbine in the basement.

In 1910, the dam was enlarged to provide more water for a stronger head, or fall to the turbine. This new dam, an earthen one, became a huge pond. It was reinforced at its outlet with concrete, and fitted with a wooden flood or sluice gate, which was lowered at night to stop the water from running out and allow the dam to refill with water. In the morning the gate was again raised sending the water back again to its site above the mill. Sometimes the force was so great some of the water at the mouth of the standpipe spilled over the edge of the hill.

When the dam was reconstructed, the old wooden standpipe was replaced with a metal pipe. This was buried in the hillside, with also the longer pipe, which was about 250 feet in length and leading to the turbine likewisc buried.

The site of the large dam remains, and its edges which years ago were raised by the steady working of horses and men, arc still visible. Grass and brush now grow over them and at one place a piece of the old sluice gate with its long turning screw lies on the ground. Nearby, the Bilhmaier Creek rushes through a narrow, rocky channel of the area at the edge of the bluff, its crashing sounds leading one to the end of the field to gaze down at the cascading water. This Bilhmaier Spring water was famous in the area's early years, and still is for its strength of flow and its purity. It has supplied the city with most of its water for years.

Jacob Bilhmaier in the 1870s, also constructed an industry of his own along the banks of the Bilhmaier Creek, which operated successfully for several years. This was a brewery and he called his product, the *Bilhmaier Beer.** A few of the heavy

*When Bilhmaier quit running the brewery a man by the name of Wolfe ran it for a year or two and then sold it to Walt Richardson who tore the brewery building down using the lumber to build a barn there.

Millstones rest in total darkness underneath the newer part of the Pataha Flour Mills, as part of the supporting structure.

Small, but strong turbine of the mill. Round opening in foreground, took the 16-inch pipe bringing water which ran the turbine. Later it was augmented with electrical power.

Some of the open gearings in alcoves throughout the mill.

Pataha Flour Mills

Detail of larger belted wheels on long shaft. The larger wheel is power, or master wheel. This is belted to the wide pulley wheel of the turbine.

Bagging station. Bag was fitted to metal frame, and flour augered down into it.

Before the sewing machines, bags were stitched at the tops with long sewing needle and twine, by hand. An 'ear' was caught at one end, twine twirled around it, then top was sewed with a few quick flips of wrist. Another ear was quickly formed at other end, and twine cut with end of needle. Long wooden lever on left was used by the operator.

timbers once used in the building still show at the site of the old brewery location, thrust up along the muddy banks of the creek.

Under John Houser's expert management, the mill's products became well known, and a constant and ever increasing volume of business poured in. John more and more was creating and developing his own speciality in flour, seeking out high gluten wheats, and experimenting in their combinations until he achieved a flour of the right high gluten content. Manufacturers of macaroni needed this type of flour, and it was also desirable in other baking needs.

The mill by 1890 had become crowded and too small, and although the millstones whirled at a smart speed under the turbine's power, and gave good results, still these millstones needed a lot of attention; constant watching for proper adjustments and periodic redressing to keep the furrows at their best with just the right edge. All this required time and was an item of expense when one of the runs must always be discontinued while its two stones were being dressed.

The use of rollers in the flour milling industry was becoming widely used by now, and all the new mills were installing them. Many of the older mills were giving up the granite millstones and enlarging their structures to hold four or five sets of rollers instead. These machines, contained in neat, compact cabinets, were equipped with corrugated steel rollers which offered a major advantage over stone milling, in that the slow grinding action of the steel rollers, each set producing a different fineness, gave a more uniform flour. Whenever rollers were installed business picked up dramatically and flour mills soon found themselves filling new, large orders. The farmer's wife preferred this new fluffy flour, for she loved the light, fine-textured bread it made.

Houser was quick to realize the great advantage in changing over to rollers and in 1891, he updated and reconstructed the mill. He enlarged the building with additions, added five roller cabinets, new bolting equipment and a more sophisticated cleaning and washing machinery. He also put in a bleaching chamber which adds little to flour except to make it whiter and fluffier. The old millstones, four of them, were removed and given an honored role beneath the mill structure, holding up the new office addition as supports.

Now, with additional space in the building, he partitioned off parts of rooms for bins to hold and store wheat. These bins holding different kinds of wheat would make his special type of flour, the *XXX Patent,* one of the most popular brands of flour ever produced in the West, sought out and ordered by hundreds of markets.

Around harvest time, many a farmer and his workers might often see a fast-moving, medium-sized figure hurrying through sections of his field gathering small bags of wheat heads. They knew it was John Houser obtaining samples of their grain.

Garfield County Gristmills

Two of the old granite millstones (there were four, or two-run) now used as supports under remodeled section. Hole in side, with one on opposite side, used for lifting with a device for that purpose, when furrows were dressed (sharpened). Pataha Flour Mills

Wooden auger, now among other mill debris.

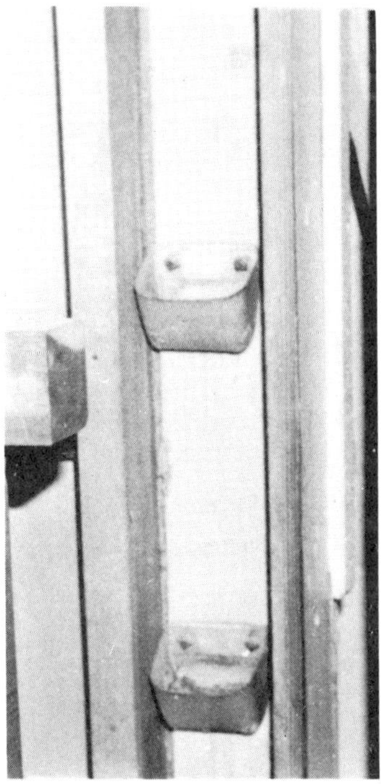

Cupped conveyor belts.

Steps in staircase are all mortised. Steps have all been worn thin, and are as slick and smooth as if waxed.

John Houser in later years.
(Courtesy of Mrs. Williams J. Houser)

The farmer would give him a friendly wave of his hand and shout a greeting hoping that when John got back to the mill and washed out that ball of dough he'd made from the wheat, there'd be plenty of gluten left. Then John most probably would make him an offer for his entire crop, and usually the farmer would sell it to him on the spot. Soon John would visit the farmer at a slack time and after a few exchanges of greeting and a hearty shake of hands, John's left forefinger would find its perennial resting place beside his nose, and the fire, always present in his steel grey eyes would spark up a little, as he'd made the farmer a good offer for his wheat. The farmer might hesitate a little, scratch his head slowly and purse up his lips, "Could you up that figure a little, John."

John's forefinger moved up and down beside his nose. He wanted that man's wheat. Best gluten content he'd found. "What do you say to, oh—to another half cent?"

The farmer nodded his head slowly a few times, then said, "The crop's yours, John."

John's standards for gluten in wheat were high, not less than 10½% gluten, which might be used in pastry flour, and preferably 17½% which would make his all purpose flour, the *XXX Patent*.

Gluten is by itself, a "tough, sticky substance obtained by washing out the starch from the wheat." Gluten also is the substance in wheat and other cereal grains that makes dough rise. Wheat which contained the higher quantities of gluten was known as a "hard" wheat and it was wheat of this variety that Houser used in his flour. The brands of wheat he liked best for his flour were Turkey Red, Forty-Fold and a variety he particularly liked called Marcus which he had to bring in from Montana.

By 1901, he took in a partner, his son William J. Houser, and now the mill took the name of *Houser and Son Flour Mill.* William had graduated from Whitman College (Academy, then) and had taken a course in business training and management at a college in San Francisco. He was by nature well suited for this business, knew it thoroughly from long association in the mill with his father, working there since a boy along with his brother Mox,* and possessing executive and business acumen similar to that of his father.

Now, with the mill under the capable hands of his son, John Houser felt the urge to indulge some of his other natural interests. He owned property at Silcott including the gristmill there, and also the water rights of the Alpowa Creek at designated points. The bars of the Snake River country were nearly all planted into fine fruit orchards in those days before dams raised the river's waters to obliterate them, and John had an orchard at Silcott near the river which he irrigated with the Alpowa Creek water. He allowed no one else the use of any of this water, to the consternation, it is said of many who just assumed water from a creek or river was free for everyone's use just because it was there. John also turned the force of some of this water through pipes he had brought from near the mill and flushed it into gravel beds to wash out gold which he was sure lay among these rocks. It is said he obtained enough gold to have a gold ring made for himself and this ring he prized very highly always wearing it proudly. The old grubstake gold miner in John Houser never completely died out.

Orders for the Houser and Son flour continued in abundance. A San Francisco baker would use no other flour, his cook's temper flaring hotly even at the suggestion

*Mox Houser bought into another local mill, owning and operating that. (See story under *Pomeroy Flour Mill*, this book.)

William J. Houser
(Courtesy Mrs. William J. Houser)

some other flour would do as well. The mill's products went to Portland, Oregon, where it was used to make macaroni, its high gluten content ideal for that purpose. It went to China, many other Asian countries, Alaska, Europe, and all parts of the United States.

When the mill was reconstructed in 1891, no other change after that time was ever made or needed in the flour mill. Much of its very early equipment remained in the mill and was still used. Some of the manufacturer's patent dates, the printing fading but still readable giving the years *1877* and *1878* show the soundness of the early wooden fittings.

Electricity came to the mill late for it wasn't really needed, although it was finally connected to the turbine to augment power and conserve water, which became necessary.

Garfield County Gristmills

Many farmers chose especially to haul their grain to Houser's mill for it was Houser's flour they wanted in return. The farmer's wheat load was dumped into a large hopper and from then on the mill took care of it, first weighing it and cleaning it. The wheat was carried in tiny buckets on a belt elevator in an enclosed metal chute, six or eight inches square, to the third floor and emptied into one of the storage bins. From these bins the grain was augered to a spout that returned it to the basement. There it was put through a washer and again moved up in buckets by way of an enclosed belt to tempering bins on the third floor. After tempering, it was moved down to the first floor to begin its feeding into the series of five rollers, each roller grinding the grain finer than the preceding one until it was reduced to flour.

Now the flour was picked up by an enclosed bucket elevator and returned to the top floor where it was run through a series of sifters. After its stay in these nervously jerking machines the flour was ready for its final grooming, going to the second floor by means of an auger where it was "bleached in sealed drums filled with chlorine gas." After bleaching the flour dropped into a hopper on the main floor where it was ready for sacking, and after sacking, shipment.

Some of the machines still have the dates of the early patents printed on them—1889 and 1890, and these machines give the appearance of being as carefully fashioned and finished as a piece of furniture, even to the varnish still clear and unmarred by time. Timbers in the old building are not weathered as one would

The printing still clear. Moline, Ill., Oct. 12, 1876; Nov. 24, 1877. The wood is sound, and fresh looking.

Net Weight, 75 Pounds
When Packed

Pataha Flour Mills
POMEROY, WASH.
Guarantees this
WHEAT MIXED FEED
to contain not less than
12 per cent Crude Protein
3 per cent Crude Fat
not more than
10 per cent Crude Fibre
6 per cent Crude Ash
12 per cent Moisture

In later years, farm feeds were an important mill product.

expect, and do not show signs of age, but appear bright and fresh with scarcely a crack anywhere in them. The ladder-like stairways, built against the walls for the quick scurrying up by busy feet, show the results of the many who trod them. Each riser, open and narrow and mortised snugly into the sides, is white and smooth from long use and worn thin in the centers. If not careful a leather sole could slip off the step, its surface as polished as if it had received a coat of wax.

John Houser passed away in 1911 at the age of 82 years. By that time William J. Houser had long taken over the management of the mill, running it under the name of *Pataha Flour Mills,* and with excellent success. William operated the mill until 1940, closing it at that time, but not for any lack of business. A new system of government supervision of small businesses brought paper work and other demands in such overwhelming, time-consuming proportions operation of the mill under these conditions became difficult.

***Pataha Flour Mills* as it looks today. Still under ownership of Houser family.**

Garfield County Gristmills

William Houser also looked after other interests which he held, such as wheat farming along the Snake River. Most of his holdings were said to have been in Asotin County. In 1903, he married Miss Emma Edwards of Lewiston who still resides in the house near the picturesque mill. This home is beside the very first house, the bungalow of John Houser and his German bride. Mrs. William J. Houser still radiates a charm and beauty that attests to the grace of an age into which a few people visiting with her are privileged to catch glimpses occasionally, and appreciate. A daughter, Frances, (Mrs. Raleigh Hampton) of Clarkston is often at her mother's home. William Houser passed away in March 1949, at the age of sixty-eight.

The gristmills in Garfield County are nearly gone, the old shell of the Colmbia Center Mill remaining and the Pataha Flour Mill intact. The Pataha Flour Mill is reminiscent of a man's work—John Houser's 'labor of love' with the basic, vital food, flour.

This was a gristmill* which followed the tide of change sweeping through the country in the late 1800s. This change was brought about by the advent of the railroads and the new mechanisms, both on the farm and in industry. It was an industrial revolution felt in force then and catching hold in tremendous power ever since, to continue onward and never to stop. This was a flour mill which not only survived that industrial change, but perhaps advanced always a little ahead of it, creating and developing in its field proportionally to the abilities of its millers and owners, John Houser and his son William Houser.

It is men like these, sternly adhering to the principles of honesty and perfection in their work who gain the lasting respect and admiration of future generations who regard those qualities in a man as important.

*For preservation purposes, listed in National Register of Historic Places.

Vanished Gristmills

GARFIELD COUNTY GRISTMILLS

ALPOWA GRISTMILL

A large haystack stands alone in a field beneath a Snake River bluff, alone except for a colony of bees close by. There is nothing spectacular about a lone haystack in a field, or a colony of bees, except this one, "Dick Ledgerwood's hay stack", rests on a historic spot—the site of an old gristmill.

The noise of the bees bears no resemblance to the long ago creeking sounds of the gristmill's waterwheel, and their droning is only a sigh compared to the grumbling of the gristmill's granite molars masticating corn or wheat for the farmer's uses. But the bees stirring and whirling around their hives bring an air of life and activity to the site of the otherwise quiet, rural setting, and long ago the gristmill was a lively, busy place.

Two streams of water flow nearby, evidenced by a thick line of willows marking their courses. They are the Stember and the Alpowa Creeks, the two joining here. Their babbling sounds have dimmed in the intervening years for the water in them has lessened, but long ago the Alpowa Creek gave up some of its ample waters to turn the waterwheel of the gristmill.

Near this Alpowa gristmill a little town once flourished. Besides the mill, there was a store or two, a post office, a schoolhouse, and a house in which the miller and his family lived. There was also the well-known "Half-way House", a stage-stop Inn half way between Pomeroy and Lewiston. Here, travelers refreshed themselves and enjoyed delicious dinners enroute from Walla Walla to Lewiston, the other stop being Pomeroy twenty-five miles away.

Families who know the Alpowa region nostalgically remember the town and the gristmill. Cliff Wilson, now at Clarkston, but raised at Alpowa, says his father "ran a fruit ranch on the bluff above the Alpowa Creek, and also dressed the stones of the Alpowa gristmill." (He used a *millpick* for this, carefully chipping away at the edges of the furrows in the stones to keep them sharp). Cliff estimates that the gristmill was built in the early 1880s "very close to that date" and can further pinpoint the location of the gristmill (since haystacks are never permanent) as "near the bridge over Stember Creek and to the left a ways."

The gristmill was built by Luther A. Stimson, a veteran millwright and miller in the Snake River country from the 1870s to the 1890s. His granddaughter, Mrs. Sydney Lee Campbell, now ninety-four, is still living,residing in La Grande, Oregon.

Through her son, (Luther's great grandson), Keith Williams of Oxnard, California, Mrs. Campbell has offered welcome information of those early days.

The Stimsons came to this region in 1876, first living at Anatone. Luther built gristmills, as well as ferries in the area. In 1877-78, he constructed the gristmill at the little town of Columbia Center in Garfield County. After this gristmill was completed he operated it for a time as its miller until they were able to find a man trained in that trade to live there and run it. In 1880-81, Stimson built a gristmill at Asotin, (*Historic Sketches of Asotin County,* Judge Kuykendall). When that mill was finished, he began the Alpowa mill in 1881.

After the Alpowa mill was completed, Luther Stimson operated it as its miller and owner for a number of years. He built a cottage very close to the gristmill and he and his family lived there.

Mrs. Campbell is the daughter of Thornton (Barney) Owsley of Pomeroy, and she grew up in Pomeroy. She remembers visiting her grandfather Luther at the Alpowa mill when she was a girl. She states, "People didn't take pictures then as freely as they do now, so the only picture I have of it is the one in my mind. I must have been under ten years old when I visited my grandfather at the gristmill there". If she were around seven or nine years old then, (her birthdate is 1883), the date of her visit would possibly have been 1890 or 1892 and at that time, Mr. Stimson was thus still operating the mill as its miller.

The Alpowa mill is believed to have been built along the same general plan as the Columbia Center mill near Pomeroy. The Alpowa mill, however, had an overshot waterwheel and the Columbia Center, an undershot.

When there was a good volume of water in a creek, an overshot wheel was usually installed because it was the more efficient, (75%). A millpond, or reservoir, was built several hundred yards away from the mill on the Alpowa Creek and this insured a steady, even flow—or head—of water to the water-wheel. A ditch brought the water to the millsite and then to the overshot wheel by way of an elevated wooden box flume.

The elevated flume, built on wooden supports, passed over the road which came to the mill, and teams and wagons, or other travelers to the mill passed underneath the flume. No doubt, they were sprinkled with water more than once, for flumes if made of wood, always leaked.

Lena Fitzsimmons who lived in that area, reminisced for the *Alpowa Pioneers:* "Water was taken out of the millpond to power the mill and was led in a ditch above the road along the hillside. It was called the millrace and the water in the millrace flowed over the millwheel (waterwheel) and energized power to grind the grains. The millstones were imported from either Austria or Yugoslavia." (John Houser of

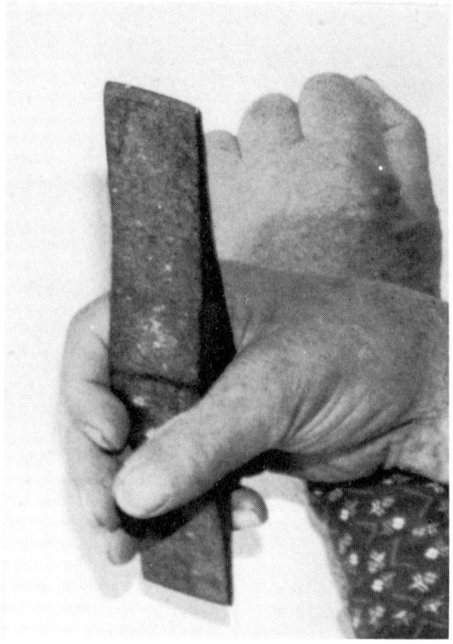

One type of millpick used for sharpening furrows of millstones. Bracing hand holding pick against wrist of other hand gives better control for tiny chipping strokes against the furrow's edges.

Pataha City, told Cliff Wilson the stones came from Czechoslovakia). "In later years, the mill ditch was used for irrigation."

The flour in the Alpowa mill was not bolted (sifted). It was simply caught in the wooden enclosure which was built around the stones. Here it was channeled down below where it either went into a storage bin, or was scooped up by hand and put into the farmer's clean sack.

The *nether,* or bottom millstone of Alpowa Mill, which is in corner of Cliff Wilson's garden in Clarkston. It is 36 inches in diameter, encircled with 11-inch metal band. Stone is in nine fitted sections.

Garfield County Gristmills

In the later 1870s, Andy Lee, a resident of the area, operated the stage stop at Alpowa, the "Half-way House", for travelers from Walla Walla to Lewiston. The Pomeroy stage stop was half-way between Walla Walla and Lewiston, and the Alpowa stage stop, was half way between Pomeroy and Lewiston. These stops were necessary to replace the teams with a fresh set. It also gave the passengers an opportunity to freshen up and rest from the long, bumpy, dust-laden trip. But best of all, it gave them a mouth-watering feast; the bounties of Mrs. Lee's deliciously laden table. Mrs. Lee was a renowned cook of those times, but she had a worthy competitor in Mrs. Pomeroy at the other stop, who was noted throughout southeastern Washington for her culinary creations. These were the days when cooking was an art, worked at and considered the most important attribute of any woman's accomplishments, the word *calories,* almost unknown.

Whether at "Lee's" or "Pum's", the roast beef was browned and cooked at just the proper temperature to hold all its juicy flavor, the vegetables were brought steaming to the table, retaining their color and slightly crisp texture of the garden. The currant jelly, freshly made, sent out wisps of tangy sweetness, a bright red mold of it quivering on a white, stoneware saucer. Bread made with whole-grain flour ground out at the gristmill only a few hours prior to its baking, was yeasty, moist and chewey, claiming attention from the other foods. Desserts were juicy apple or berry pies, or creamy puddings piled high in a dish. It was worth thinking up excuses to go to Walla Walla or Lewiston, just to be able to tuck napkins under chins and fork savory morsels into hungry, eager mouths.

The Half-way House was said to be located, "one quarter mile up the gulch northwest from the main Alpowa Creek", on property originally owned by D.S. King who had "squatted on the land." On it was a good-sized log house, believed to have been built by Thomas Grant who was a nephew of General Grant, the noted general then. It was in this building that the stage inn was established. Lena Fitzsimmons remembered it in a brief description: "It nestled below and near a huge bluff of moss-covered rocks. There was an orchard of fruits and berries growing there, watered from Stember Creek".

The gristmill provided a means of entertainment for many of the youngsters, for the water in the tail race, after its use with the water wheel, formed a little pool where the children and some of the adults enjoyed skating in the winter, and sometimes swimming in the summer, if it got deep enough. Picnics were popular and could be held most anywhere, perhaps in some little glen near the mill on the banks of the Stember of Alpowa. Here, wild roses, serviceberry bushes, and waxy syringa bloomed in abundance and formed natural garden spots, with birds adding an extra attraction with their trills and flourishes.

Luther A. Stimson and his wife, Betsy Lamb. Photo taken in Colfax at the time Mr. Stimson was operating a ferry boat.

Notice Luther's capable looking hands.

(Picture courtesy Mrs. Sydney L. Campbell, granddaughter of Luther A. Stimson)

Garfield County Gristmills

Mortised and tenoned joints, secured with long wooden pin. Mortising was the hollowing out of opening in beam, into which was fitted the tenon, or tongue, of another beam.

Some of the farmers who used the gristmill might have been L.B. and Wm. B. Hawley who jointly farmed and raised cattle on 1200 acres, J.A. Howard, who lived three miles north of Alpowa, H.C. Hutchinson, two miles west of Kelly's bar who was a wheat farmer and fruit grower, Wm. P. Keffer, two and a half miles north of Alpowa, Jay Lynch, Lewis E. Malone, Sidney Dresser, D.S. King, Felix Landkammer, Dave Smylie, John Shaner, John Henley, Tom Shangle, Roland Weissenfels, Sam Smith, Ed Whitman, Wm. Freeman, Fuller and Walker, who owned a large sheep ranch here. There were others farming and raising stock: B.C. Montgomery, John Bushnell, Wm. and John Galloway, James Manners, L.R. and W.B. Hawley, the Flocks, J.N. Boggan, Pinkhams, Shumakers, Woods, Robinson, Favor, Whiton, Trescott, Welch, West, Skinner, Thomas Boggs, Wormall, Romaine, Wamsley, Pine, Water, Vennergarholz.

Farmers who lived along the high ridges also used the mill: the Hawleys, Howards, Dudley Strains, H.C. Hutchinson, Wm. Keffer, Jay Lynch, Lewis Malone, Thomas Ruark, Harmon Scott, N.A. Wheeler who was the post master for a time and also took over the toll road. John Cook, J.A. Howard, I. Harrington, J. Lynch, Robert Bracken, John Henry, Columbus Walker, Joseph Harris, Otto Munch, Wm. Holland, L.A. Porter, John Alfrey and Martin Weiss, and others.

Luther Stimson stopped operating the mill in the early 1890s, and turned to constructing and operating ferry boats. He operated a ferry at Asotin and later moved it down near the junction of Wawawai Creek on the Snake. Stimson devised a convenience new to ferry boat operators in pioneer times. He constructed "ferry houses" on both sides of the river and in these shelters the ferryman could wait for the bell to jingle, a signal Stimson installed which would signify a party was waiting on the opposite shore for the ferry boat.

Stimson sold the ferry in 1899 to his son, Luther Curt Stimson and his son-in-law, J.C. Dodson who kept and ran the business a year before they sold it. Stimson's date of death is not exactly known by Mrs. Campbell, but she believes it must have been around 1899 or 1900, for she never remembers seeing him after that.

Luther A. Stimson was born November 30, 1832, near Des Moines, Iowa. His wife, Betsy M. Lamb was born in October 1832. On November 4, 1856, Luther and Betsy were married in Porter, Michigan. They came to the vicinity of Garfield County in the early 1870s by way of Wyoming and Idaho and settled in the Asotin country, at Anatone.

Luther's training in millwrighting and milling was in the East and a well-grounded one. He must have built a number of gristmills there before coming West. Happily, there is visible evidence of his skill as a millwright, for the 'shell' of the old Columbia Center mill which he built, still stands, and the workmanship which went into it brings admiration and respect for this man's craftmanship. There were no nails used in the interior construction, and the massive beams are all mortised and tenoned, and further secured by long wooden pins. These joints are as strong and secure now as they were when first fashioned, one hundred years ago. All Stimson's work was of this superior quality, and would still endure today it not torn down, such as was the Alpowa mill in 1925.

The Luther Stimson family was a large one, as families were then, with three sons and six girls. The daughter Harriet Eliza Stimson, who married Barney Owsley, was Mrs. Campbell's mother. The Stimson girls must have been very pretty as well as capable young women, and there is just a hint of the romance between Barney and Harriet who was seventeen at the time of their marriage, and whom Barney always called "Lize". In later years he related his experiences for the *Pioneer Life in Washington:* "In 1876, Luther Stimson and his family entered the Alpowa country settling at Anatone. It was there I met his daughter Harriet, when I went into the Grand Ronde Valley to raise cattle in 1878. We were married in 1879 and spent a year on Joseph Creek right across from Old Chief Joseph's ranch."

Keith Williams, who writes for his mother Mrs. Campbell, remembers his grandmother, Harriet Owsley who lived in Pomeroy, "Grandma Owsley and I were great buddies when I was little and we had lots of fun. She had a great sense of humor, was kind and gentle and taught me to play the harmonica so that we could play duets. She died at the age of 92 in Pomeroy." Her death was in 1954 and Barney's was in 1938.

Close-up of mortised and tenoned joint, which is pinned with long, wooden, pointed wedge. Head of pin is visible.

Luther and his wife Betsy are both thought to be buried at Pullman but the graves which were unmarked, have never been located definitely. Mrs. Campbell, Sydney, and her sister Bess (later Mrs. W.B. Kuykendall), spent an entire day when they were at college, trying to locate the Stimson graves, but without success.

After Stimson left the Alpowa mill around 1893, Andy Lee purchased the gristmill from him, and Andy Lee and his two sons, Bob and Barrik operated and ran it for several years. Lena Fitzsimmons remarked: "He did custom milling for many farmers who brought corn and wheat to the mill to be ground into cornmeal and graham flour. Mr. Lee charged the customary toll (usually one-eighth) as a fee for the work."

The Alpowa gristmill stopped running in the early 1900s. Even before that time, the little town had begun to dwindle, the fate of most small villages throughout the region. The stage stop was no longer needed for the smoke from the locomotives' stacks was marking a new and faster path for people who preferred to trade in larger towns, such as Clarkston and Lewiston, rather than the inadequate town of Alpowa. Farmers were taking their grain to the newer mills equipped with the roller machinery, which produced much fluffier and whiter flour than the whole grain product ground out between granite millstones.

Small gristmills throughout the country such as the Alpowa gristmill, the Columbia Center mill, and similar ones in Walla Walla county, Whitman county and other places, which did not enlarge and install the roller system found

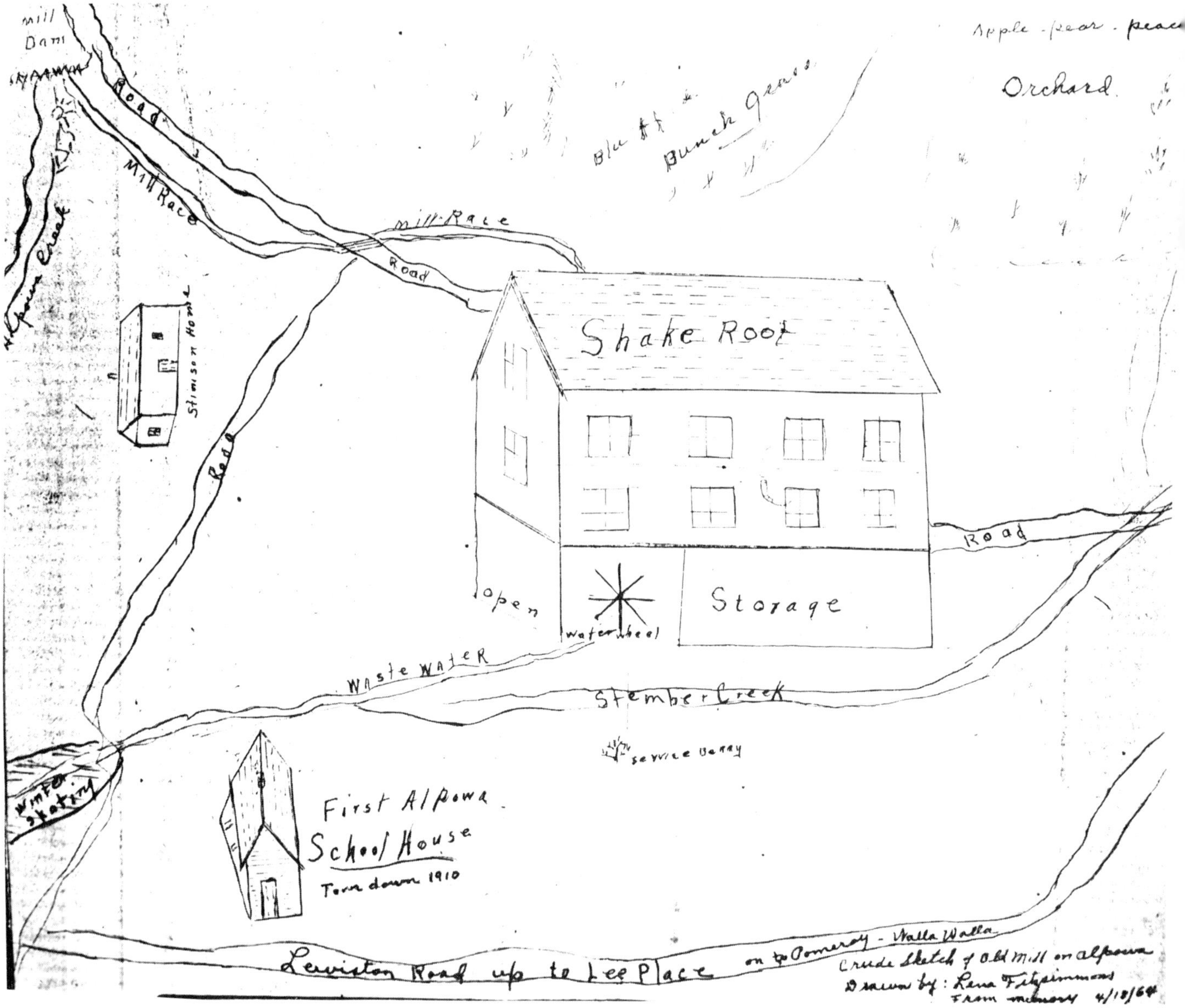

Lena Fitzsimmons' crude sketch of the Alpowa Gristmill and surrounding area. Mrs. Fitzsimmons lived at Alpowa. The sketch was made in her later years from memory.

Courtesy Garfield County Historical Society

themselves in want of customers. It was no longer profitable for them to stay open to operate for only a few families who still preferred the old-time, wholesome and healthful flour of the old gristmills.

The Alpowa Mill stood idle several years, the machinery inside it rusting. Field dust mingled with the decaying residues of flour dust, cob-webbed along stout wooden beams, or settled into piles at floor corners. Finally, in 1925, the old mill was torn down. The line of the mill ditch still shows plainly on the hillside but the rest of the area is now planted to hay. Bees take pollen from the alfalfa blossoms and hay stacks often dot the field. Occasionally one of them rests on the exact spot where once the old gristmill stood.

Vanished Gristmills

GARFIELD COUNTY GRISTMILLS

GOULD CITY

On September 8, 1888, a small notice appeared in the East Washingtonian stating that "Santo Brothers are erecting a chop mill at the forks of the Deadman and are almost ready to begin running it. Unfortunately, the flume of this mill blew down but should be back up shortly."

The mill which the Santo Brothers built, began life as a gristmill to grind flour. They installed an overshot wheel, with an elevated flume for carrying the water, by way of a ditch, from the North Deadman. The mill was originally built for Alec Hender and constructed on his property. After it was finished, Mr. Hender operated it first and Roy Toyer then ran it. The Santo Brothers also ran the mill, it is said.

It was equipped with two small granite millstones, each about 18 inches in diameter. These had been made in an eastern firm, and were shipped 'around the horn' to Walla Walla.

Gould City chop mill powered by water from North Deadman Creek. Young Harold Graham in baby buggy. Car is either Stevenson's or Rauche's. (Courtesy Garfield County Historical Society.)

Garfield County Gristmills

One of the two millstones first used in the Gould City gristmill. Stones are small.

The mill manufactured flour for only a short time. Other larger flour mills were so near, the farmers preferred to take their grain either to Pomeroy or to Pataha City. The Gould City Mill was then converted to a chop mill, and this operated very successfully grinding chop feed for the farmers.

In 1898 or 1899, Marion Graham, grandfather of the present Harold Graham now living on the home farm, bought the Hender property which adjoined Graham's farm, and this included the mill. Marion Graham then took over the mill and continued operating it for several years.

Harold gives some interesting sidelights on the mill. He stated that a "flour breaker", or roller, was bought from John Houser of the Pataha flour mills who did not plan to use it in his mill, and it was installed in the Graham mill to use for chopping feed. The millstones were removed. They now are in the Harold Graham yard and each used as a "planter" for flowers growing in their centers.

The Graham family operated the mill along with their farm work. When the men were too busy especially during harvest season, Harold says that the family ran it. Marion Graham had six children, and Agnes Graham, twelve years old, and her brother Frank, fourteen years old, ran the mill mainly by themselves. One of the men would start the machinery going for them, and they would operate it during the day while the men were working. Agnes even learned to sew the sacks, and became as adept at this task as her father.

The mill was a two and a half-story, wooden frame building. When it was changed to a chop mill, a turbine was installed to replace the water wheel. The turbine is still in place in the farm yard. The pipe and flume to the turbine is said to have had a thirty-foot drop, which gave it a strong force.

Earl Victor who was brought up in the Mayview country, up beyond Gould City, says he remembers the country grocery store—usually called general merchandise store, at Gould City. When he and his brother had been to Pomeroy with their dad for some business there—the boys riding in the jolting hack behind the farm team—they'd always stop at the Gould City store. Here their dad would buy them each an old-fashioned bottle of pop. He remembers how good that tasted, so cool and tangy. He also recalled that in the early 1900s, there was not one single bridge between Mayview and Pomeroy, and they would have to cross the several streams and creeks at the lowest places. Sometimes this was quite a risk, especially during high water when the creek bed might be washed out under the swift water.

Besides the grocery store at Gould City, there was a post office, and a blacksmith shop run by a man named Castle.

The mill ditch is still visible for it was used for irrigation through the years, and the turbine is still standing, part of it submerged. Even the old mill's door is in use as a gate-door on the barnyard fence, and it still has the Santos Brothers' name on it, clearly printed in black letters.

Mill's turbine still stands in farm yard.

The old mill door. A tree's shadow crosses it.

The mill ditch.

The little town which was there, a tiny one, is completely erased now. Harold points to a low place that is covered with a concrete bridge, and recalls to a visitor that "the store stood at that spot."

The chop mill was used well up into the 1920s, and beyond, finally being torn down in 1955.

Some of the early farmers who might have used the chop mill were the Blachley family, the Victors, Ole Hannas, McGreevy, Luke Kidwell, John Lynn, the Whitmores, John Burr, and perhaps the McPhersons, the Oscar Millers, Stuarts, and the Hayes family, and the Whitworths, the Schneblys, the Charles James, George V. Black, and others, especially farmers from the Ping and Mayview districts.

The chop mill at Gould City was a flour mill for only a short time. Another chop mill operated in Pomeroy. This was on Pataha Creek at the west end of town, said to have been located where Helke's Trailor Court now stands. This chop mill was run by Sam Nevell, and operated partly by steam and partly by electricity.

Asotin County Gristmills

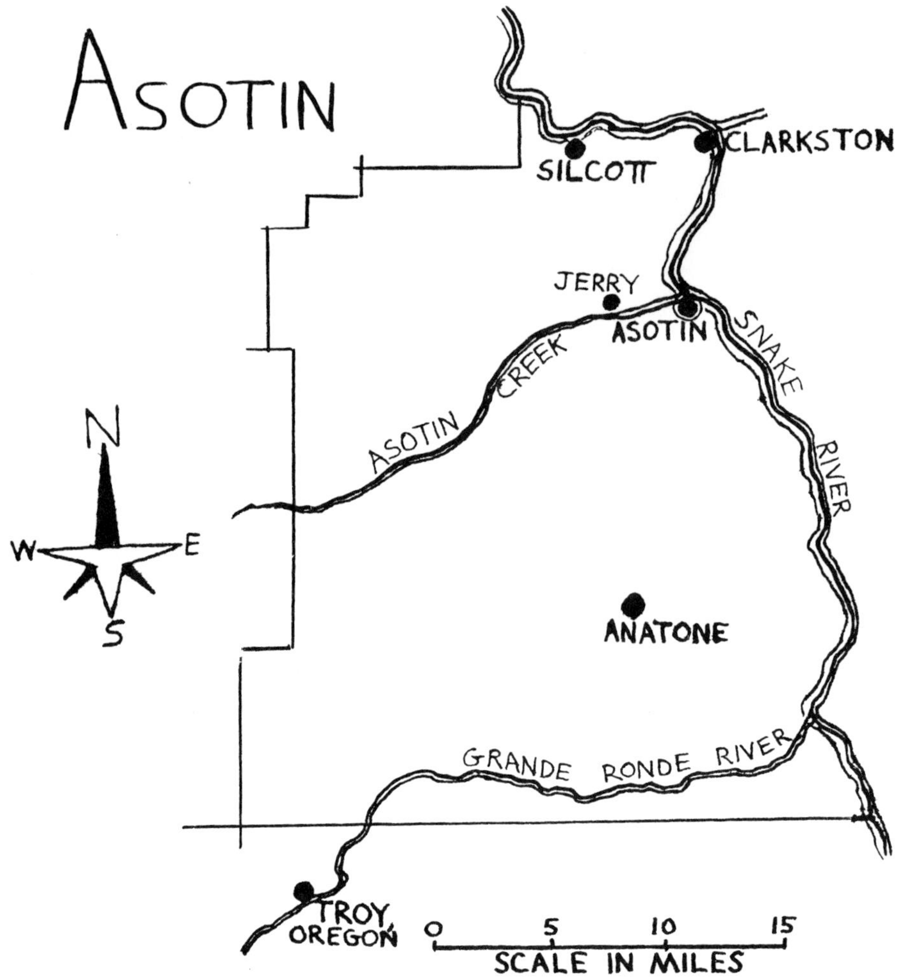

GRISTMILLS IN ASOTIN COUNTY

Silcott
Asotin Flour Mills (Jerry Milling Co.)
Anatone Flour Mills
Jerry Mill at Jerry

Asotin County Gristmills

SILCOTT

ON THE WEATHERED slopes of Silcott's high bluffs, the saga of the Snake River country is written in sand and wind. It is told in the rocks which trace jagged edges all along the bases of the bluffs, and chanted everywhere in the wide rivers and streams, and outbursts of sparkling springs.

Years ago, this region at the mouth of Alpowa Creek was a perfect home for Indians. It remained so until the covetous eyes of the whites peeked over the bluffs, followed soon by the whites themselves seeping into the region, reaching their long, grasping fingers everywhere.

The Nez Perce Indian tribe lived here. The waters of the Alpowa Creek emptied into the Snake in a low, almost tropical-like flat area. Timothy was chief of the Nez Perce tribe and Red Wolf was chief of all the Nez Perce people in the region. These Indians were friendly to the whites, and when Spalding came to live among them and work with them, they helped him find the location at Lapwai.

Spalding, working tirelessly with these Indians, established a "branch mission" at the home of the Alpowa Indians. Spalding gave them much help in matters besides religious teachings, helping them with agriculture and planting apple trees on Red Wolf's farm in 1837. He also built a crude gristmill there for the Indians' use.

This gristmill was a very small one, patterned after the one Spalding used at Lapwai. It was also like the one at the Whitman mission at Waiilatpu, for Marcus Whitman had purchased a gristmill like Spalding's.

The gristmill on the Alpowa can best be described as a tubmill. It had a pair of millstones, cut out of native rock of the region. The millstones had a vertical shaft about five feet from the ground which was connected to a horizontal water wheel in an open tub-like container. Water, which turned the wheel, poured down into it from a height of about six feet, coming to it by way of a wooden trough bringing water which had been collected in a small dam or reservoir.

Vanished Gristmills

On December 8, 1894, the *Asotin Sentinel* published the following item: "H.H. Spalding of Almota, (the missionary Spalding's son), presented the State Historical Society at Tacoma with the first millstone ever used in Washington. It was ground out of a piece of granite along in the forties (1840) and used by the Indians near Alpowa to grind the grain they were taught to raise by Spalding's father (Missionary Spalding) and his colleague, Dr. Whitman. Alpowa, where the stone was used, was the home of such noted chiefs as Red Wolf and Timothy, and is located in Asotin county on the Snake River. The millstone was hand-hewn out of a piece of gray granite, a foot thick and has a round hole seven or eight inches in diameter in the center".

Timothy, and his wife Tina who was a sister of Chief Joseph, had a daughter, Jane. Jane, under the guidance and influence of the Spaldings, received an education and learned and followed most of the ways of the white people patterned after those at the mission at Lapwai. Jane was sought after in marriage by many chiefs, but Timothy declined to follow the Indian custom of selecting a husband for his daughter, telling them that his daughter would be permitted to marry whom she chose.

Jane married a half-caste Nez Perce by whom she had a son who was drowned. Her husband also died and later Jane married John Silcott, a white man, who came there in 1860 to erect government buildings at Fort Lapwai. When his contract was completed he remained in the area and built and operated a ferry across the Clearwater River not far from its mouth. He and Jane were a happily married couple.

In later years, 1885, the town which had developed here on the Alpowa, took the name, *Silcott,* in memory of Jane Silcott.

When the white men glimpsed this serene Alpowa site, they wanted it. By 1870 or before, nearly all the Indians were gone from this area, gradually pushed out by the white man claiming one site after another. One by one the band of one hundred fifty or two hundred Indians crept silently away.

On Red Wolf's property, the old orchard still stood producing fruit abundantly. It was called the "Old Mission Orchard", for it was the one Spalding had started for the Indians. Timothy had 160 acres near it, which he had acquired under the homestead act. Under this act, an Indian could not dispose of his property until a period of twenty-five years had elapsed. Although he received countless tempting offers for the property, Timothy held it for the required time before selling it to one of the white settlers.

Timothy remained in the region until his death in 1890 at the age of 90 years. A few other Indians remained: Indian Moses, who "lived above the Weldon place," and Indian Henry Powaukie, who lived at the mouth of a small stream which entered the

Asotin County Gristmills

Timothy was an early convert to the Christian religion under Missionary Rev. Henry H. Spalding. Timothy and other Nez Perce Indians helped Spalding establish his mission at Lapwai. Later, Spalding established a small Indian church at Alpowa, a "Second mission." He also taught the Indians agriculture and put up a small primitive gristmill at Alpowa, similar to the one at Lapwai, for the Indians' use.

(Illus. courtesy Glen Adams. Drawing by Randall Johnson.)

Timothy's, and his daughter, Jane Silcott's grave at Beachview Park in Clarkston, moved from shallow graves near the river in 1915.

Timothy's Inscription: "First Nez Perce Christian convert and preacher. Led Col. E.J. Steptoe's troops to safety and ferried them across river at Alpowa when they faced annihilation by hostile Indians, May 25, 1858."

Jane's Inscription: "Chief Timothy's daughter, wife of John Silcott. Guided Capt. E.D. Pierce around Nez Perce reservation to gold diggings near Pierce, Idaho, in 1860, thus initiating gold mining in Idaho."

Alpowa. Henry was much liked and respected by every one. His sister, Lucy, was known among the whites as "The Princess," and was a beautiful woman. They were related to Chief Joseph and Chief Red Wolf.

Types was perhaps the best known. He, a convert to Christianity as the others were, refused to drink any alcoholic beverages and scorned anyone who did. When once offered a small bottle of medicine for an illness, he, smelling the alcohol in it, pushed it away with his hand saying, "Whiskey, no good, me no take."

In 1882, the quiet little settlement of a few white men and families which had developed here at the mouth of the Alpowa, sparked to life. Garfield county had just been established in 1881 and a new county seat was to be selected. A number of little towns were grooming themselves as candidates to attract votes for the prize position, and Alpowa decided to enter the race.

E.D. Miner, a surveyor, laid out a forty-acre townsite of twenty blocks, and a number of business establishments took foothold. White's ferry was already operating. Now, a post office was established, a schoolhouse built, stores put in, a big warehouse with a capacity for 100,000 bushels of grain built, a saloon put up. Most important of all, a large, well-equipped gristmill was built by Ray Woodworth and D.M. White, and became a significant part of the town.

David M. Mohler, now living on the old Red Wolf place, had on this property an old irrigation ditch. This ran along the edge of the property and down to the former Timothy homestead. This ditch, according to the Indians, had been built in 1838 to bring water to irrigate the orchard.

In 1882, Ray Woodworth who lived on the property next to the Timothy place, bought from the Mohler family the right to flume the water past the Mohler place and use it to bring water for the operation of the gristmill.

The gristmill "was erected about two hundred yards west of the Silcott schoolhouse, close to the bluff and about opposite the residence of the Michael family." It was operated by water from the Alpowa Creek, carried in a wooden box-flume three-quarters of a mile in length from the water collected in a small dam.

A new name, *Alpowa City* was given to the spruced-up town. But it's aspirations for fame as a county seat were short-lived, for it, along with many other similarly hopeful towns, was eliminated. The contest had finally narrowed down to a battle between Pomeroy and Pataha City, with Pomeroy eventually the victor.

The Ray Woodworth mill, as people called it, operated for four years and would have continued on for many years longer for it was a successful one, used widely by the farmers in the region, but in July, 1886, it caught fire and burned down. It was said to have contained the finest equipment available at that time, which could have been roller machinery operated with the water-powered turbine. The mill building was a two and a half-story structure.

It was said that at one time John Houser, outstanding miller and owner of the Pataha Flouring Mills, owned the Silcott (Alpowa) mill and ran it for a while.

Alpowa City died down again after its brief, exuberant moment. It came to life again in 1885 under the new name of *Silcott* given it. This was in remembrance of Timothy's Jane. Seth Reynolds took over the post office and Tom Robinson and his brother Lewis took over the Old Mission Orchard property. They operated the old fruit orchard for many years for it still was productive. Other fine orchards grew in the district as the mild climate was ideal for the raising of fruit crops. The flour mill was not rebuilt after it burned.

The bars and banks of the Alpowa and the Snake River at Silcott were often placer-mined for gold. People could be seen working diligently with pan or shovel along the gravel and sand banks. These were the days when many Chinese were imported to work on the railroad and there would often be as many as two hundred Chinese laboring to the point of exhaustion for the gold. It is said some earned around a dollar a day at this.

Gold mining like this was hard work and the gold was difficult to come by because it was 'flour' gold, and very difficult to separate from the sand.

All the landmarks of the Ray Woodworth mill at Silcott were removed when the army engineers put in the Lower Granite dam in 1974. The highway was moved and relocated, said to be "directly over the Michael place." The location of the old schoolhouse was adjacent to the highway, "easily found by those who remember where it stood."

The entire area of what was Silcott is now under water from the effects of the dam. Silcott's life as a town was a relatively short one under that name, but its location and its background was a long and ancient one. Its first original Indian name, Alpowa, means "place of rest", and perhaps describes the site best.

Like the Indians who lived in this region and left forever when the white men took over their home land, so now the tribal site is also gone, buried forever under water.

The old gristmill that Spalding built vanished ages ago, and the gristmill built in 1882 which served the farmers in that area for such a short span of time, is also vanished, now only a memory.

Vanished Gristmills

ASOTIN COUNTY GRISTMILLS

ASOTIN FLOUR MILLS

Asotin Roller Mill
Jerry Milling Company

The Asotin Flour Mill, which later became the Jerry Milling Company, and finally the Marvel Mercantile Company, is one of the most interesting of all early flour mills in Eastern Washington, mainly because of its age and durability. It undoubtedly is the original mill built in Asotin in 1881 by Luther A. Stimson.

Its life has been a long, but steady and purposeful one of more than ninety-six years, with little or no dramatics involved. Like a faithful farm animal, it served the farmers and community well, grinding out flour and feeds. Just as it looked and seemed at its worst, decrepit and run down, its entire appearance and function was changed and restored to an attraction. The mill stands in its original location, a notable gristmill example and a point of pride to the town and the people of Asotin.

Asotin itself is unusual. Besides always having been a desirable living area because of its temperate climate, it has enjoyed many commercial advantages as a shipping point because of its location on the Snake River. It is situated about nine miles below Clarkston at a point at which the Asotin Creek empties into the Snake, and has an elevation of about 760 feet. This area offered everything the early Indians as well as the first white people into the region could wish in a home.

One of the first of the many white men coming here, was Jerry Maguire. He chose a location along Asotin Creek in 1867 about three miles from its mouth, and it is notable that this later took the name of "Jerry".

By 1867, most of the Indians who originally inhabited these lands had left, feeling the force of the white man pushing them out. Some remained. One was Tom Hi, or California Tom, as he was called, who stayed in the region. California Tom was a smart Indian with lots of the white man's savvy. He quickly caught on to the white man's money-making operation of packing goods into the mining camps, and did likewise. Later, when the white men took up farms and began building on their ranches, Tom improved his place faster than any of them, erecting the first frame barn in Asotin County.

The white people settled at one end of the Asotin bar, calling their community *Assotin City*. In 1878, Alexander Sumpter, a surveyor, platted it into a townsite with it developing by 1880. After the post office was established, a gristmill was started.

Asotin County Gristmills

The gristmill was built by Luther A. Stimson and Frank Curtis, and finished in 1881. It was run by water from Asotin Creek collected in a dam. A ditch, partly of wood, ran alongside the hill carrying the water to the gristmill. At a point above the mill, a flume took the water to the wheel. This is believed to have been an overshot wheel at first and a few years later changed to a turbine, with the flume along the hillside elongated and changed then also. When the mill was built, it had one run of granite stone and turned out 50 barrels of flour a day.

Almost from the beginning of Assotin City's existence as a town, another community developed as its rival at the opposite end of the bar, calling itself *Asotin.* For many people, Asotin was a more favorable location that Assotin City. T.M. Schank platted it in 1882 and offered lots so the people would be interested in getting the town under way. Much rivalry and differences grew between the two communities but it was obvious there could not be two towns practically in the same spot with almost the same name. Although conflicts arose, the towns under some firm legislative persuasion, finally merged in 1887, the surviving one being *Asotin.*

The gristmill built in 1881 continued operating during all this time, grinding wheat for the farmers in the area. Some people living there in those days who might have used the gristmill were Jackson O'Keefe, William Reed, Louise and Alexander Reed, Edward Baumeister, W.J. Clemens, Wm. Critchfield, J. Swain, John Weissenfels, J.E. Bushell, D.B. Pettijohn, McAlphen, S.T. Jones, W.H. Wood, S.S. Rogers, George Bailey, Charles Lyon, Thompson, George Fancher, James Hopwood, Lige Jones, Thomas Page, Noble Henry, Samuel Warren, James Thorn, Lile D. Dittemore, Al Short, Wm. Robers, P. Brown, N. Ausman, H.J. Allen, M. Mitchell, J. Mitchie, Bingham, J. Wann, John Winson, C.C. Goodwin, S.F. Bennet, J.N. Rice, P.C. Brown, H.C. Fulton, F. Waldrip, L.W. Tate, Steward Leonard, Charles Brantner, A.W. Chapman, A.R. Powell, C.W. Caywood, P.W. Dodson, M.T. Johnson, James Carter, F. Werseheid, J. Tate, O.T. Green, J. Meador, O. Montgomeroy, D. Bingham, W. Cooper, Joseph King, J. Oatheimer, W.L. Cook, J.W. Jones, Wm. Farrish.

By 1888, the gristmill had another partner replacing Stimson who was busy building other gristmills. An old directory states that Frank Curtis and L.G. Braden were running the gristmill under the name of *Asotin Flour Mills.* These owners improved the mill, adding another run of stone, changing the loading platform and replacing the water wheel with a turbine. The mill still used water to power it. It was at this time the wooden flume was elongated, still following the hillside's curve and extending to Clemens addition to take water there for irrigation. Above the mill, a sharp cutoff of concrete was built, connecting to a flume or standpipe which dropped to the mill bringing water to the turbine. The output of the mill was increased from 50 barrels a day to 125 barrels.

Vanished Gristmills

Mill about as it looked when built in 1881 at Asotin by Luther A. Stimson.
(Picture courtesy Robert Weatherly, Asotin, Wash.)

Picture of Asotin Flour Mills, looking North. Two open water flumes like long necklaces follow along against sides of bluff. Higher flume took water to Clarkston, bottom one above the mill, to land near Clemans Addition. Notice short flume at right angle to long one, where it drops down to the mill. Picture taken around 1897 before mill remodeled.

(Courtesy Robert Weatherly, Asotin, Wash.)

Asotin County Gristmills

Mill teams and wagons loaded with wheat, beside Asotin Flour Mills before third floor was added to it.

Grain haulers to the Asotin Flour Mill. These pictures taken in 1890s before mill was remodeled and made larger. Photos courtesy Joseph Barron

About this time, at a much higher point all along the top side of the bluff, another wooden box flume was built paralleling the lower one, both of them resembling two long necklaces encircling the bluff's neck. The upper flume carried water to Clarkston for irrigation.

In 1896 or before, the mill was taken over by Charles Isecke, a well known business man originally coming from Anatone in 1888. He was also a miller and millwright. One historical source states he built a mill at Asotin, but reliable records show that he bought into the Asotin Flour Mills — the original mill built in 1881 — and possibly rebuilt it. Isecke installed the roller process equipment and enlarged the mill. The third story was possibly added at this time and the loading platform extended. An early deed indicates Isecke was an owner of the Asotin mill and that his partner was Bradford G. Church. The mill now was called *Asotin Roller Mills.*

About 1890-1900, Mr. Vollmer, a banker and business man in Lewiston, bought into the Asotin Roller Mills, along with A.C. Barron, an experienced miller from Crawfordsville, Indiana.

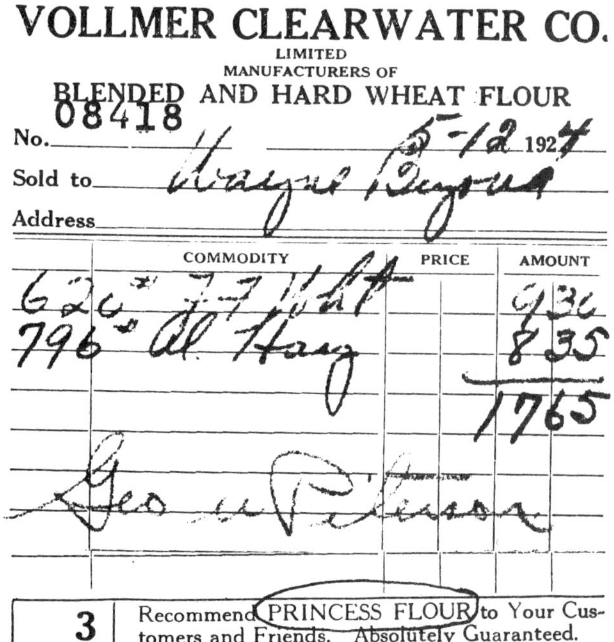
VOLLMER CLEARWATER CO.
LIMITED
MANUFACTURERS OF
BLENDED AND HARD WHEAT FLOUR
No. 08418 5-12 1924
Sold to
Address

COMMODITY	PRICE	AMOUNT
626# ... Wht		9 30
796# Al Hay		8 35
		17 65

3 Recommend PRINCESS FLOUR to Your Customers and Friends. Absolutely Guaranteed.
169689 Mfd. by United Autographic Register Co., Chicago

Vollmer Clearwater Company owners in the 1920s until 1934.

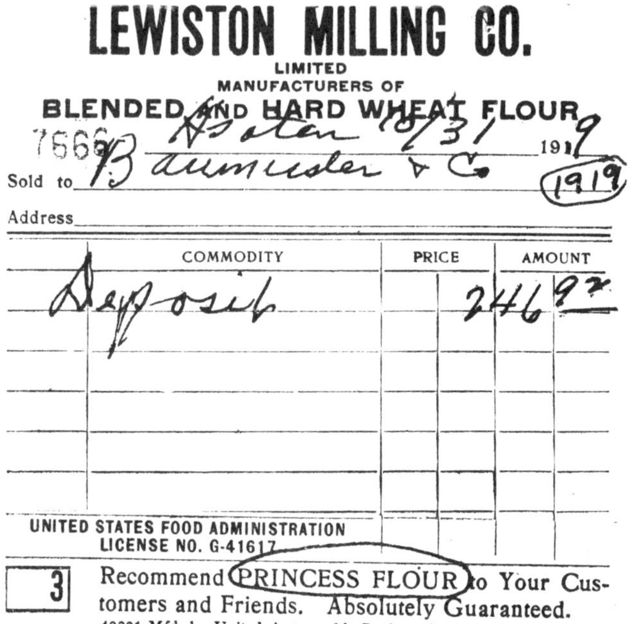
LEWISTON MILLING CO.
LIMITED
MANUFACTURERS OF
BLENDED AND HARD WHEAT FLOUR
7666 Asotin 10/31 1919
Sold to Baumeister & Co
Address

COMMODITY	PRICE	AMOUNT
Deposit		246 92

UNITED STATES FOOD ADMINISTRATION
LICENSE NO. G-41617
3 Recommend PRINCESS FLOUR to Your Customers and Friends. Absolutely Guaranteed.
49391 Mfd. by United Autographic Register Co., Chicago

Lewiston Milling Company owners in early 1900s.

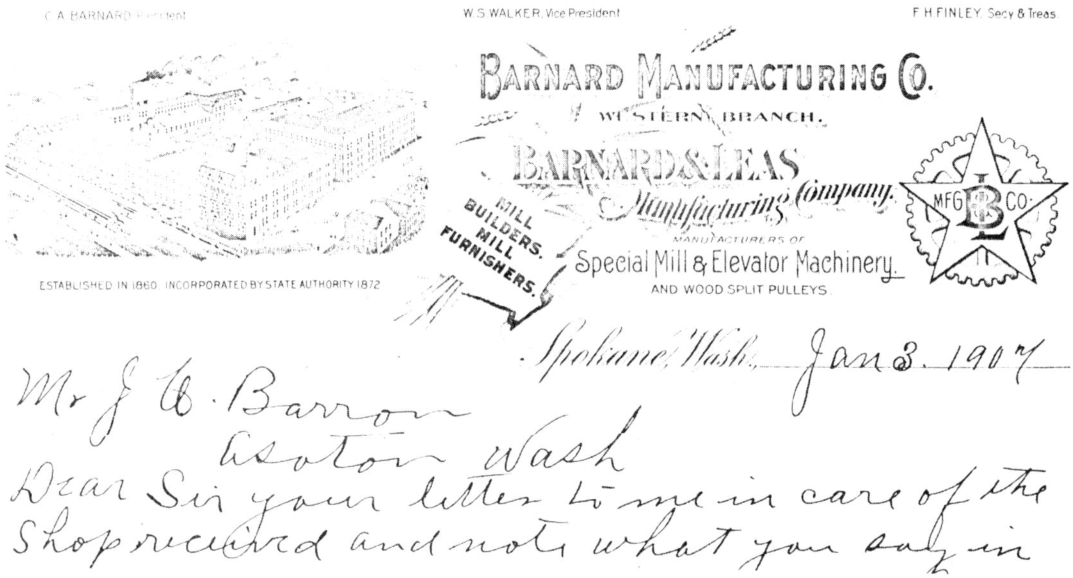
C.A. BARNARD, President   W.S. WALKER, Vice President   F.H. FINLEY, Secy & Treas.

BARNARD MANUFACTURING CO.
WESTERN BRANCH.
BARNARD & LEAS
Manufacturing Company.
MILL BUILDERS. MILL FURNISHERS.
MANUFACTURERS OF
Special Mill & Elevator Machinery
AND WOOD SPLIT PULLEYS
MFG B L CO.
ESTABLISHED IN 1860. INCORPORATED BY STATE AUTHORITY 1872

Spokane, Wash., Jan 3. 1907

Mr J C Barron
Asotin Wash
Dear Sir your letter to me in care of the
Shop received and note what you say in

Hand written business letters and an old deed and mortgage are still in possession of J.C. Barron's son Joseph Barron at Oakesdale, giving information which helps piece together the Asotin mill's ownership and history. The Barnard Mfg. Co. installed the roller equipment in many Eastern Washington flour mills. this was a Spokane branch. They also helped many millers locate other mills for purchasing, or operating.

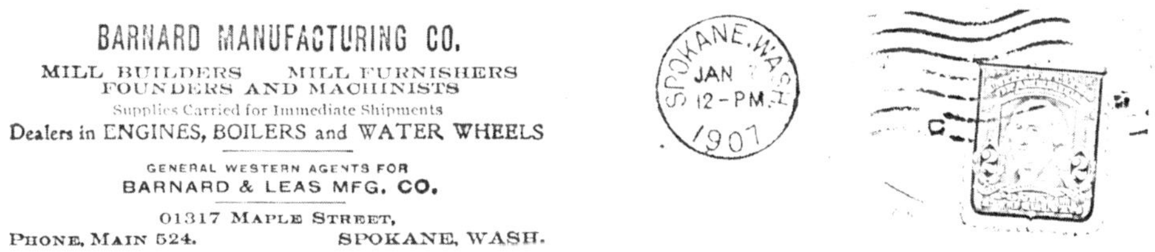
BARNARD MANUFACTURING CO.
MILL BUILDERS   MILL FURNISHERS
FOUNDERS AND MACHINISTS
Supplies Carried for Immediate Shipments
Dealers in ENGINES, BOILERS and WATER WHEELS
GENERAL WESTERN AGENTS FOR
BARNARD & LEAS MFG. CO.
01317 MAPLE STREET,
PHONE, MAIN 524.   SPOKANE, WASH.

SPOKANE, WASH.
JAN
12-PM
1907

Envelope of Barnard Manufacturing Company. Note two-cent stamp, and postmark.

(Courtesy of Joseph Barron)

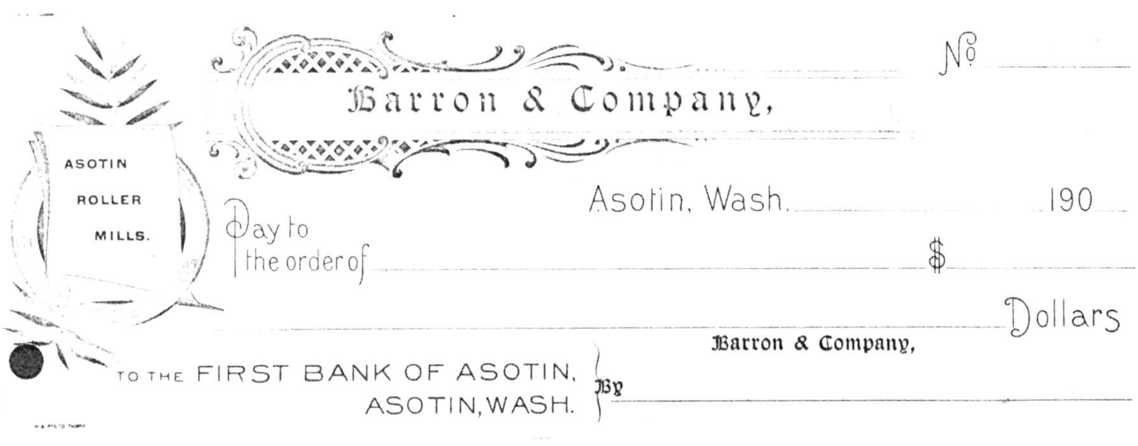
No.

Barron & Company,

ASOTIN ROLLER MILLS.

Asotin, Wash. 190

Pay to the order of $

Dollars

Barron & Company,

TO THE FIRST BANK OF ASOTIN,
ASOTIN, WASH.
By

Courtesy Joseph Barron

Vanished Gristmills

In 1903, A.C. Barron sold his interest in the Asotin Roller Mills to his brother Joseph C. Barron, also an experienced miller from Indiana. J.C. Barron now took charge with Mr. Vollmer also part owner, the mill operating under the name *Asotin Roller Mills,* operated by *Barron & Company.* Joseph C. Barron remained with the Asotin Roller Mill until 1906, at which time he bought into the flour mill at Oakesdale. Mr. Vollmer then bought out Mr. Barron's interest in the Asotin mill, May 15, 1906.

After 1906, the Asotin Roller Mills was under the operation of the Lewiston Milling Company with Vollmer the main owner, and in 1919, it came under the ownership of Vollmer-Clearwater Company of Lewiston.

In later years, 1934, the Vollmer Clearwater Company sold the Asotin Roller Mills to three men. These gentlemen were D. Alonzo Silver of Jerry, F.H. Stephenson and Homar Robeson of Asotin.

An interesting sidelight is worth mentioning, which shows a slight relationship of the Asotin mill to other mills—the mill at Anatone and the mill at Jerry, explained as follows:

For several years in the early 1900s, D. Alonzo Silver owned and operated the gristmill at Anatone, but in 1919, he closed it down, moving to Jerry to build a gristmill there. The Anatone mill was dismantled, and Silver took the roller machinery from it and placed it in the new mill at Jerry, three miles up Asotin Creek from Asotin.

After Mr. Silver bought into the Asotin Mill in 1934, he closed down the Jerry Mill, discontinuing its operation to give his full time to the mill at Asotin.

When the three men took possession of the Asotin Roller Mills in 1936, they found the mill's machinery "had been all stripped out". Whether this arrangement was part of the purchasing agreement or not is not mentioned in information from the Robeson brothers.

But again, the old Anatone milling machinery came to the rescue, for now that roller equipment was removed from the Jerry Mill and taken to Asotin and installed into the mill there. The Asotin mill was also given a new, but familiar name, *Jerry Milling Company.*

The owners of the Asotin Jerry Milling Company, now turned to a new source of power, although the turbine was still intact, not having been removed when the other equipment was taken out. They installed a diesel motor. The old mill was again in top working condition, and it immediately began grinding out flour, producing large quantities of it continuously from then on, until the end of World War II.

Asotin County Gristmills

View of town of Asotin in early 1890s and 1900. Bridge is across Asotin Creek. Large building on right is Mr. Baumeister's store. Upper floor of this building often used as a meeting hall or recreation center.

Home and three children of J.C. Barron, part owner and operator of Asotin Flour Mill in 1903. Children are Fred, Grace and Alma. Grace, center, later became a teacher in the Pomeroy Public school system.
Note water well at left, with its wooden keg used as a water pail.

Photos courtesy Joseph Barron

By the 1940s or before, Hormar Robeson's two sons, Jack and Gerald were old enough to enter the business. But flour milling by this time had turned unprofitable for the small mills, since it was fast becoming an industry for the larger corporations. Another business opportunity was opening up for which the mill was well suited, and that was the important one of processing livestock feeds, grinding grains and other feeds and compacting them into pellets and other chewable sizes which could be sacked and sold to farmers for their stock. The mill now stopped grinding flour and took on this new enterprise, changing over to the working of a chop mill. The diesel power plant was removed—taken to Elk City to generate electricity there—and the Asotin Jerry Mill was now wired to run with electricity. The mill continued in this productive and profitable business from then on, running at full capacity up into the middle 1970s, processing livestock feeds.

Although the appearance of the old mill by this time presented a worn, down-at-the-heel look, a closer scrutiny would have revealed the structure was not nearly in

Author's sketch of the Asotin Flour Mill, renamed *Jerry Milling Company* in 1936. It continued in operation until 1974, when it came under new ownership.

as bad shape as might have appeared at first glance. The underfootings had taken a bad beating from many floods of the creek, and needed some work done there, but otherwise the inside beams were still stout and sturdy, the bracings of joints as secure as when first fitted together. The timbers still held straight and firm. Decayed flour dust might have settled in a few stark corners, or edged some thick beams with cob-webby lace, but even this could not hide the fact that strong boards and planks still supported heavy machinery and storage loads without the slightest give.

On the outside, the grey, weathered clapboard siding remained secure. The porch and platform needed attention. The main structure, however, was sound. The Robeson brothers who must practically have grown up in the mill, stoutly came to the defense of their building whenever the occasion warranted, with, "That old structure may look worn and weathered, but it's a long way from being dead."

The Coast Trading Company of Portland evidently came to the same conclusion when they examined it in 1974, for they purchased the old mill that December.

Shortly after that, the Marvels; Bill and his father Clifford in Asotin, took over the building and turned it into a mercantile company. After this strong company purchased it, the mill building underwent unbelievable and miraculous changes, not only in appearance but in function as well. First, came a general sprucing up inside and out, the repairing of underfootings, and other structural parts, such as the porch and loading platform. They put on a new roof and made improvements so the building could accommodate merchandise which would be put into it. Last, they put on a coat or two of bright yellow paint, which finished its transformation.

This old mill had not only claimed a long and useful life from 1881 to the 1970s, but beginning in the later 1970s was embarking on a new career. It was unquestionably good for another hundred years or more, barring some unforseen catastrophe.

It is one of the very few water powered mills built in Eastern Washington in the early days—this one, 1881,—which continuously operated throughout all the years of its existence, and is still standing.

In many ways, the gristmill's development followed that of the town and region, and the lives of the people living there all through those years. It's world reflected the community's world.

Asotin and the surrounding country-side is a beautiful region, and still a desirable place to live. The town rests peacefully by the Snake River. It is sheltered by steep, dramatic looking, high bluffs, up through which concrete highways now point the way to and beyond the fertile wheat fields lying on their tops.

Vanished Gristmills

FROM THIS

Old *Jerry Milling Company*, originally *Asotin Flour Mills* and then *Asotin Roller Mills*, and built by Luther A. Stimson in 1881. Remodeled in late 1890s or early or early 1900s by Charles Isecke.
(Courtesy Robert Weatherly, Asotin, Wn.)

TO THIS

Marvel Mercantile Company Rejuvenated building of the *Jerry Milling Company*.

Asotin County Gristmills

ANATONE FLOUR MILL

In 1878, a Daniel McIvor and his wife, are said to have been in the Anatone region living in a cabin which they built and constructed to serve partly as a little store.

That same year, Charles Isecke came into this locality and purchased all of McIvor's interests. Isecke began a little settlement, enlarging and running the store along with a post office. The little town of Anatone was well started by the 1890s but it was not platted until 1901.

In the 1890s, a flour and feed mill was built. By then a number of farmers living in the wide prairie regions saw the necessity for such a business, and banded together to build one. Among the many who worked in getting the mill underway were McIntosh and Sangster who are said to have been instrumental in building the mill and operating it. The mill was powered by a steam boiler outfit, for no major stream flowed through the area which might have furnished water power.

Anatone Flour Mill. Picture taken 1908. **(Courtesy Robert Weatherly, Asotin)**

The town of Anatone as pictured in 1908. Located on the wide prairie lands above the town of Asotin and the Snake River. Lies in the very southeastern tip of Washington, with Idaho and Oregon state lines nearby. The higher peak showing is *Big Butte*, el. 5009 feet. (Both pictures courtesy Robert Weatherly of Asotin.)

Main street of Anatone in 1908. Could almost pass for a set from a Hollywood movie with the lonely figure of a woman walking through the town in the middle of the street. Portion of gristmill shows at right of picture.

Buildings in Anatone, identified by number.

1. Bowersock Livery Stable
2. Commercial Hotel
3. E.V. Bowercock's Home
4. Bradley and Mitchell Store
5. Clemans Store, with Social Hall Upstairs
6. Methodist Parsonage
7. Methodist Church
8. Crow's Blacksmith Shop
9. Anatone Flour Mill
10. Bert Crow's Hotel
11. Wheat Storage Warehouses
12. Stuckey's Home (proprietor of flour mill)
13. Chris Halvorson Butcher Shop
14. Chris and Ole Halvorson Store
15. Anatone Public School

Asotin County Gristmills

The farmers brought in their grain to the mill and traded it for flour, bran, rolled grain, or any other product the mill could produce.

In the early 1900s, D.A. Silver is said to have owned and operated the mill at Anatone for several years. By 1919, the mill no longer was producing as well as earlier, and he closed it down. He took out the machinery in this mill and moved it to the town of Jerry on Asotin Creek. There, Silver constructed a new mill into which he put the machinery from the Anatone mill. Silver called the new mill at Jerry, the *Jerry Mill*.

Farmers in the Anatone country who could have used the Anatone mill might have been Clemens, Bracken, Bradley, Toops, Benedick, Sangster, Barnes, Whiton, Taplin, Herring, Chapman, Lane, Carter, Farrish, as well as many others.

Vanished Gristmills

ASOTIN COUNTY GRISTMILLS

JERRY MILL at Jerry

Jerry Maguire was one of the first white men to come to the Asotin area in 1867. He chose a location along the Asotin Creek about three miles from the point at which it enters the Snake River. In later years, this location took his name and has always since been called "Jerry".

In 1919, D. Alonzo Silver "moved from Anatone where he operated a mill, to the community of Jerry." Here, Mr. Silver started building a flour mill. Its location was said to be "beneath the bluff across the creek at the Cloverland junction". Silver took the roller machinery and some of the lumber from the Anatone mill which he had closed down, and brought it to the new Jerry mill, using the lumber in the new mill's construction and installing the machinery in it. The mill was operated by electricity.

Although the mills in this book are mainly water-powered gristmills, this electrically-powered flour mill is included, for it holds a sort of blood-relationship to the Jerry Mill in Asotin. When Mr. Silver joined F.H. Stephenson and Hormar Robeson to found the *Jerry Milling Company* in Asotin in 1934, Mr. Silver closed down the Jerry mill in Jerry. He took out its machinery—which had originally been in the Anatone mill—and hauled it to Asotin, installing it in the Jerry mill at Asotin. This Jerry mill in Asotin, was originally the Asotin Flour Mills, which was originally the old Stimson mill built in Assotin City, in 1881.

By now, most readers are completely baffled, as this writer was at one time. Anyone can be excused for saying, "Oh, I thought the Jerry Mill was at Jerry."

The answer quickly comes, "That's right, it was."

From then follows more explanation of the relationships of the three mills to each other—same owner, same machinery, same name—from Anatone, to Jerry, to Asotin, and it all is quite clear—almost.

Jerry Mill. Picture taken in 1927.

Truck used here is World War I surplus truck alongside Jerry Mill in1921 or 1922.

(Photos courtesy Bob Weatherly)

Vanished Gristmills

Big Bend & Central Washington Gristmills

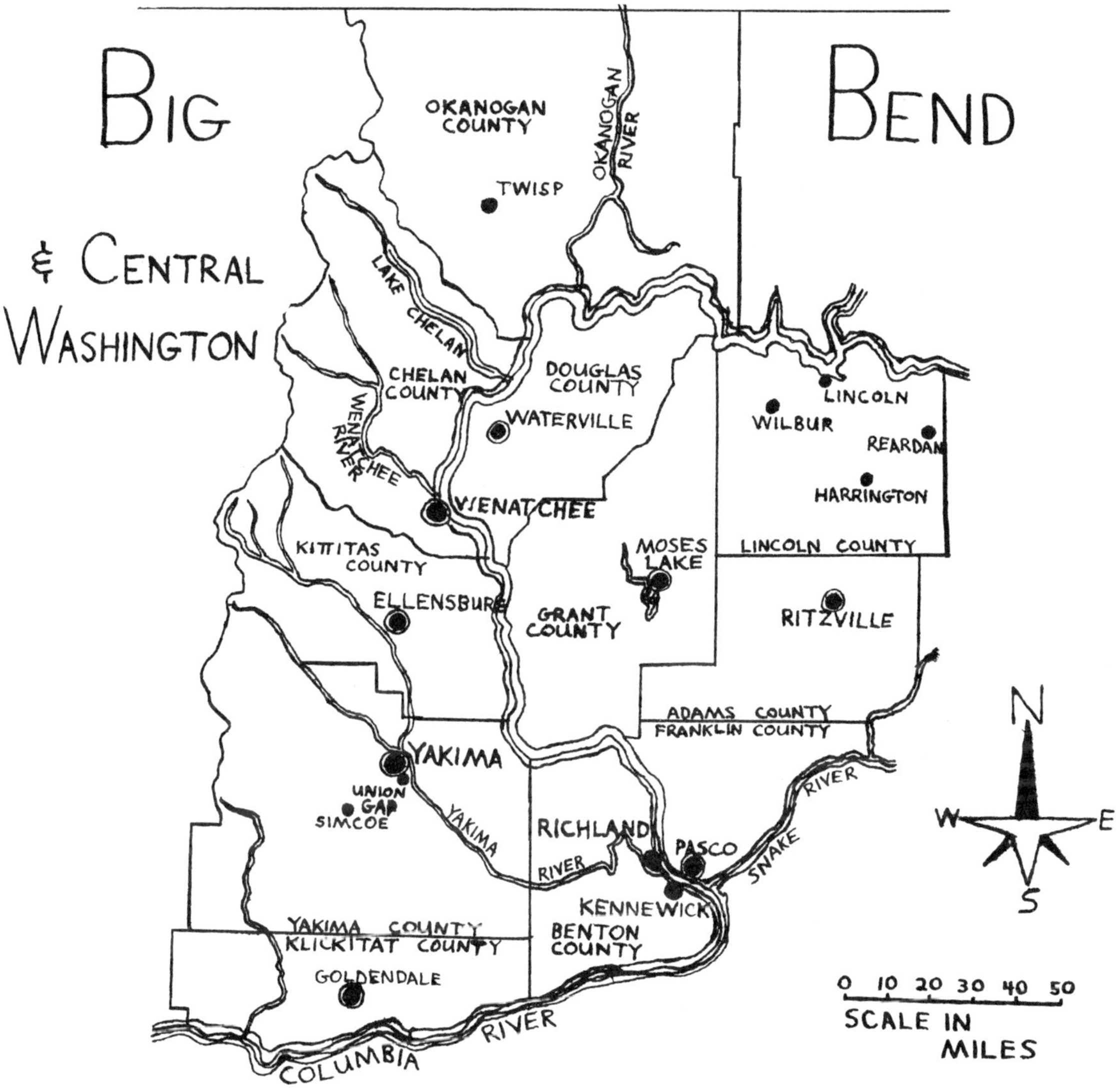

GRISTMILLS IN BIG BEND & CENTRAL WASHINGTON

Sprague Mill, Lincoln Co.
Reardan Mill, Lincoln
Holmes Spur Mill, Kittitas
Brick Mill, Kittitas
Damman Mill, Kittitas
North Star Mill, Okanogan
Klickitat Mill, Klickitat, Goldendale
Wenatchee Mill, Chelan
Waterville Milling Co., Douglas

Wilbur Mill, Lincoln
Harrington Mill, Lincoln
Ritzville Mills, Adams
Thorp Gristmill, Kittitas
Tjossen Mill, Kittitas
Sanders Mill, Kittitas
City Mills, Kittitas
Nashland Gristmill, Unknown
Schanno, Yakima at Union Gap

Simcoe Gristmill, Indian Res. in Yakima

Vanished Gristmills

LINCOLN COUNTY GRISTMILLS

SPRAGUE

BIG BEND country's first flour mill was built in 1884 by a man named David Gunning. Mr. Gunning had come to this area the year before, 1883, from Oregon, interested in building a gristmill here. In early 1884, he began the organization needed to get the gristmill going. The lumber came from sawmills near Cheney, and the machinery was purchased from a firm in Portland. Gunning built the mill as a combination millstone and roller process mill. This would satisfy nearly every grinding need of the farmers for the mill could grind the grains between the millstones into whole grain flours or meals, and stock feed as well. The rollers would produce a finer grade of flour than the millstones, and the flour would be bolted which would not only help produce a whiter, fluffier flour but would sift out the bran and shorts, which the farmers needed.

Before the mill was started, the farmers' cooperation was sought. They pledged to furnish 50,000 bushels of wheat a year to keep the gristmill going so it would operate at a profit to the owner and miller.

The farmers more than met their quota after the mill began operating, bringing in a great deal more wheat than was pledged. This meant that all the farmers in this vast country brought their wheat to the mill for grinding. This region covered all the areas between the Columbia River and Rock Creek, a distance of many miles. It is said that a large number of the farmers would come with their wagons of wheat, and camp in the mill area for three to six days, waiting to get their grinding done for the winter supply of flour.

The mill cost $10,000 to build. It was a three story building 36 X 48 feet in dimensions, of wooden construction with a stone and brick foundation. It is said to have been located "on a spot where a branch of the Edwall Grain Growers is now located." It had a capacity of 75 barrels of flour a day, and its turbine engine was thirty-five horse power.

Water for the mill came from the Colville Lake in Sprague, through its outlet, Crab Creek, and near the natural falls of this creek.

Big Bend & Central Washington Gristmills

The Davenport Tribune of April 1920, mentions that in 1887, a Mr. Chapman settled on the land on which this mill was built, and through which the Crab Creek flowed. Mr. Chapman was so intrigued with the falls of this creek, that he named them "Minnie Falls", supposedly after an old sweetheart by that name. The name of the falls has remained the same over the period of time, and this spot is a favorite one for fishermen.

On August 28, 1894, a half interest in the mill was sold to H.W. Boone of Spokane.

A Spokesman-Review article of January 23, 1895, mentions that the capacity of the roller mill was doubled that year, 1895.

But in 1896, a little more than ten years after it was built, the gristmill was destroyed by fire, at a loss of $7,500 to the owner. The mill was not rebuilt.

In later years a newspaper account said, "There is nothing now standing to mark the spot where the gristmill once stood, a relic of pioneer days, except the falls, Minnie Falls, tumbling over the rocky cliff."

Vanished Gristmills

LINCOLN COUNTY GRISTMILLS

There is very little information available for the following early gristmills. They are mentioned here, however, for their importance in the 1890s, when they were built and used.

REARDAN FLOUR MILL

In 1890 Reardan had a 150 barrel capacity flour mill operating, according to a Spokesman-Review item, January 7, 1890.

This mill may have been destroyed by fire for the Spokesman reports on June 20, 1899, "Washington Grain and Milling Co. is building a $50,000 flour mill with a capacity of 300 barrels of flour a day.

WILBUR

Named for Samuel Wilbur Condit, widely known as *Wild Goose Bill.*

In 1890, a 150 barrel per day flour mill was built there.

RITZVILLE

In 1897 had a roller mill, "which operated day and night", and this mill was enlarged to 300 barrels a day in 1899.

HARRINGTON

A corporation was formed in 1899, called Harrington Milling Co. to build a flour mill.

Big Bend & Central Washington Gristmills

KITTITAS COUNTY GRISTMILLS

THORP GRISTMILL

The Thorp gristmill was built in the year, 1877, by J.L. Mills. It was run by water power and the remains of this old mill are said still to be standing in the 1970s.

The water to run the gristmill came from the Mills Canal, built around 1877 by J.L. Mills, who built the Thorp gristmill and also the sawmill adjacent to it at about the same time. The huge overshot water wheel gave power to both these mills, with the sawmill having furnished the lumber for the gristmill. Later, a Mr. Kendall is said to have owned the gristmill.

The early farmers of the region hauled their grain to the mill receiving a number of services. They could pay to have it cleaned for seed grain, or have it ground into flour, or they could have it chopped into livestock feed for their farm animals. Thorp, itself, a community of about 200 busy people, was an important trade center for the area, with the many dairy farmers using this town as a shipping point for their products.

The Thorp gristmill ran for many years as it was when built, but in the early 1900s, Mr. Mills had generators installed in the sawmill with the thought of providing electricity for the town and all the residents of Thorp. Mr. Mills was the owner of the light plant and he operated this plant, but mainly at night for during the day he was much too busy taking care of his many other duties; the business of the gristmill and the sawmill, and the work on his farm. The people of Thorp realizing this, tried to adjust to his schedule and cooperate with him in their needs for more and more electricity.

Mr. Mills finally conceded, however, he should spend Monday and Tuesday mornings operating the electricity plant. These were the weekly wash and ironing days, and probably baking days, too, for the housewives in the town. When a town's entire female population firmly insisted it needed electricity on at least two mornings, it is very likely Mr. Mills agreed with them and decided it best to turn on the plant for them then.

There were other problems for the light plant which couldn't be solved as easily as the housewife's request for electricity on Mondays and Tuesdays, and these headaches were more often than not caused by the elements. In winter, ice clogged the river, or froze the canal, abruptly stopping the water supply, or high winds tore down the lines which must be strung up again before service could be continued. Countless such blows often brought the little plant "to its knees".

Vanished Gristmills

Thorp's population of more than 200 people made it a busy, active community. Besides its industries already mentioned, it operated many logging activities. Thorp was not far from the larger town, Ellensburg, so it felt itself well situated both as a rural community town, and also as a suburb to a larger town.

The old mill still stands, though now unused. The people in the area are justly proud of the old landmark, for it supplied many older generations with flour. People living there, some in their nineties, recall how important the mill was to them, remembering that it furnished them with all their flour at one time.

Thorp gristmill in Kittitas County built in the 1870s. This old mill not only ground wheat into flour for farmers, but cleaned their wheat seed, or ground grain into livestock feed for the farm animals.

Big Bend & Central Washington Gristmills

KITTITAS COUNTY GRISTMILLS

TJOSSEM MILL

The *Tjossem Mill* and its millpond were built in 1879, on Wilson Creek about four miles southeast of Ellensburg. This had one run of millstones. In the 1880s, the mill was changed to a combination roller and millstone process. The rollers were installed to produce a finer flour, but the stones were kept to do custom grinding for the farmers who preferred that type of flour.

For each 60 pounds of grain brought in, the farmer was given 35 pounds of flour and 10 pounds of bran, leaving the mill 15 pounds as its fee for milling.

The mill burned down in 1900 and was not rebuilt.

SANDERS MILL

In 1888, C.A. Sanders established a gristmill on Wilson Creek, two miles northeast of Ellensburg. It began as a mill using the millstones, but soon installed the rollers converting to the new method of processing flour. After the rollers were installed the mill had a capacity of 90 barrels of flour a day. This gave the mill the distinction of being the largest mill in the county then. In 1889, it caught fire and burned to the ground.

CITY MILLS

In 1887, the *City Mills* was built in Ellensburg, by the builders Shoudy and Tjossem. This mill was equipped at the start with the roller machinery and had a capacity of 100 barrels a day. After the first Tjossem Mill burned in 1900, part of the machinery from it was brought to the *City Mills* and added with new, up-to-date machinery that was installed.

After operating the mill for one year, the two men dissolved their partnership, with Mr. Shoudy taking over the mill and taking his son in with him as a partner.

Mr. Tjossem bought another well-producing mill, and this mill was named *Holmes Spur Mill,* located two and a half miles southeast of Ellensburg. Mr. Tjossem took in his son, Albert, as his partner.

The Holmes Spur Mill is said to have burned the next year, 1890, but it was immediately rebuilt, being replaced by a much better mill equipped with the new roller equipment. It was now able to produce 100 barrels a day.

Vanished Gristmills

KITTITAS COUNTY GRISTMILLS

BRICK MILL

The *Brick Mill* was the name given the gristmill built in 1875 by the Canady brothers on Wilson Creek about five miles northeast of Ellensburg. The mill ran by water power from the creek and had one run of granite millstones.

In the 1880s, it was replaced by a new brick building and into this mill went the newest roller equipment then obtainable, giving the mill a capacity of 75 barrels of flour daily.

Much of the Canady Brothers' flour went to the northern mines and a large quantity of it was also sold to the Indians in the region.

The property was purchased in the early 1900s by W.T. Morrison, but the mill did not run very long after those years, remaining idle for a number of years. Other mills in the region were producing flour on a larger scale by that time, and taking the business from such mills as the Brick Mill and other similar mills throughout the farming villages.

DAMMAN MILL

In 1879, a mill built and operated by J.C. Damman was erected on the west side of the river across from Ellensburg. Millstones were the method of grinding in this mill, one run being used. In the 1880s, the millstones were replaced by the roller equipment. Nearly all mills by that time were being built with roller equipment, or being remodeled to include them.

As far as the Damman Mill's success was concerned, it is said the Northern Pacific railroad's location interferred with it, for it ran between the river and Ellensburg, and thus the gristmill's use was discontinued.

Big Bend & Central Washington Gristmills

OKANOGAN COUNTY GRISTMILL

NORTH STAR MILL

A mill was built at Twisp by Hoarce Hutchinson in the 1870s. For this mill, Hutchinson bought rights to use the water of the James Mill pond for the *North Star* gristmill. No other information comes to light about this mill.

KLICKITAT COUNTY GRISTMILL

KLICKITAT MILL

A mill was built in 1878 by Thomas Johnson at Goldendale, Washington, and called the *Klickitat Mill.*

CHELAN COUNTY GRISTMILLS

WENATCHEE

J. Shotwell erected a gristmill on the Wenatchee River, Wenatchee, Wash., in 1893. One of the largest in the region, a roller mill with a 200 barrel capacity. In 1910 it was totally destroyed by fire, cause unknown. The mill was rebuilt.

WATERVILLE

A mill built here in 1891, was called *Waterville Milling Company.*

NASHLAND

"A gristmill erected by Colonel Nash near the Badger Mountains, 1887. During the winter men spent much time getting out timber for the proposed mill. Persons visiting Big Bend expressed the opinion it would be the granary of Washington Territory." Spokesman Review, 8-24-87.

Vanished Gristmills

YAKIMA COUNTY GRISTMILLS

SCHANNO MILL

There was one early mill built in Yakima county and this was at Yakima City, which was the name first given to Union Gap.

The Union Gap gristmill was built in 1869, and at first had an overshot water wheel powered by water from Wide Hollow Creek. This millsite at that time belonged to Frederick V. Barker in 1873.

Later a turbine replaced the water wheel, still using the water from Wide Hollow Creek. After the water had done its work in the turbine, it was returned to the creek via the tail race.

This gristmill stood for many years and is believed still in place, taken over by Larry Burmaster. The structure has had a varied existence, having been rebuilt once and moved twice. After the mill burned in 1930, it was rebuilt, although it was damaged by a big flood in 1931.

SIMCOE MILL

The Puget Sound *"Business Directory and Guide to Washington Territory, 1872"* prints a *Table of Gristmills.* In this list one gristmill is mentioned for Yakima County, built by the government on the Yakima Reservation. At that time, it was the only gristmill listed for Yakima County. It had one run of stone and a capacity of 30 barrels of flour per day. The mill was water-powered, and called the *Simcoe Mill.*

Big Bend & Central Washington Gristmills

PLEASE NOTE:

The author is aware that the foregoing material on gristmills in Central Washington can be considered lacking and inadequate in content, as compared with the many other gristmill write-ups in this book.

This has not been an oversight, nor has it been done intentionally to slight the fine, numerous, water-powered gristmills which were in the Central Washington areas before 1900.

Very little printed material or other information on these mills seems to have been available when the search for that much needed data was being made. Undoubtedly, much knowledge about them does exist, especially in the regions where they once stood. Some people still living in those areas may possibly know about them and could give important details.

Should there be anyone reading this, who knows about any of the gristmills in Central Washington or the Big Bend, and would like to give details and information which might be used in another edition, or in an addendum for this book, such help would be greatly appreciated.

Please write to the author in care of the publisher of this book. Thank you.

Florence E. Sherfey

Miscellaneous Gristmills

A few other gristmills, close to but not in the vicinity of Eastern Washington, are included here mainly for their importance at the time of their use.

MISCELLANEOUS GRISTMILLS

TROY, OREGON, gristmill

Troy, Oregon, is about eight miles over the Washington line from Asotin County, and so close that many of the Grouse Flat people in Washington trade at Troy.

Troy is in timber country and is a hunting center for that popular sport which so many enjoy especially in the fall of the year. In the 1890s a gristmill was built there, said to have been started by Alonzo Silver and his brother Weldon Silver. The mill does not operate now, but a little town still remains.

The mill was run by water power, with water piped to the mill from the Wenaha River, then called the Little Salmon.

POST FALLS, IDAHO

A *Cable Mill* at Post Falls built in 1890s. (Photograph of mill included.)

VANCOUVER, WASHINGTON

The *Cedar Creek*, or *Cedar Gap Creek* gristmill located ten miles out of Woodland near Vancouver, Washington. (A photograph of this mill included.)

Miscellaneous Gristmills

Cable Mill at Post Falls built in 1890s. Long cable extends to mill at right to run it. In 1897 turned out 2700 barrels of flour per month. Operated for several years by the Darts.

Courtesy George T. Dart

Vanished Gristmills

Cable Mill and Sawmill on Post Falls. (Courtesy Washington Water Power)

Bridge at left crosses Wenaha River. Mill at top of store building in center. Pictures were taken in 1912. (Pictures courtesy Robert Weatherly)

Post office and store on left, school in back, Mill at right.

Miscellaneous Gristmills

Cedar Gap Creek Mill, ten miles out of Woodland near Vancouver, Washington. Built in 1876 by George W. Woodham family, and A.C. Reid. Lumber cut in Reid's sawmill one-half mile down stream from site. Original water wheel was removed in 1886 and replaced with a turbine. Vancouver Historical Society is restoring mill. (Courtesy Spokane Public Library)

Florence E. Sherfey

Glossary

Auger: Tool for boring holes in wood; also, when imbedded in a bin of wheat carries wheat to another location by its continuous turning.

Bolting: Sifting of flour into grades of finer flour.

Bran: The skin or husk of grains, separated by sifting.

Breast Wheel: A water power wheel in which water enters buckets at middle of wheel.

Buhrstone: A siliceous rock used for millstones; also called *burrstone,* and *millstone.*

Conveyor: An endless moving belt with cup receptacles.

Distillery: A place where distilling, especially of spirituous liquors takes place.

Eye: An opening in top millstone to allow grain to enter for grinding, and into which shaft is locked for turning the stone.

Fall: The action of water by gravity to point of impact on water wheel.

Flogging: Old-time term for sifting. A beating or sifting of meal or flour.

Flume: A narrow, usually man-made ditch or pathway through which stream runs.

Flume Box: Also called box-flume; a man-made trough to carry stream's water.

Fore Bay: A deep receptacle or dam at end of mill race, from which water is admitted to wheel.

Grist: Grain that is to be ground.

Gristmill: A mill to which grain, or grist, is carried or taken to be ground into flour.

Head: Distance water falls until its impact against wheel.

Hopper: A funnel-shaped box above millstones holding grain which flows into stones for grinding.

Lantern Gear: A gear resembling a lantern with vertical dowels.

Meal: Ground grain which contains both husk and flour.

Middlings: Any of the inner coarser particles of ground wheat mingled with flour. A mediocre flour.

Overshot Wheel: A water power wheel in which water enters buckets at top.

Penstock: A pipe, also flume or millrace, but usually used to bring water direct to the turbine, and placed vertically or in an inclined position for greater water force.

Rynd (also *rind*): A piece of iron running across eye, or hole of upper millstone, serving to support the stone.

Glossary

Shoe: A device used to control grain from pouring directly into eye of upper stone. Usually a narrow trough under outlet of the hopper.

Shorts: A by-product of wheat milling that consists of bran, germ, and coarse meal.

Sifting: See bolting, flogging.

Sluice Gate: A watergate or floodgate, sometimes used at opening of dam, or in a flume or ditch, which controls the amount of water flow to the water wheel.

Tailrace: The channel, or swath of water leading away from water wheel, after use.

Toll: A portion of grain taken by the miller as compensation, or fee, for grinding.

Undershot Wheel: A water power wheel in which water enters the buckets at the bottom.

Tub Mill: A simple mill. Small millstones were attached by a vertical shaft to a horizontal wheel and rotated at speed of wheel. The wheel was propelled by the force of a moving stream, or falling water.

Florence E. Sherfey

Bibliography

Alpha Study Club, *Oakesdale Memories*, Oakesdale, WA. (n.d.)

Bal, Peggy, *Fairchild, Heritage of the Spokane Plains*, n.p. 1976

Barker, Dr. B.B., *Letters of Dr. John McLoughlin*, Portland, Ore. 1948

Becher, Edmund T., *Spokane Corona*, Eras and Empires, Spokane, WA., 1974

Bond, Rowland, *Early Birds in the Northwest*, Spokane, WA, 1972

Drury, Clifford M., *A Tepee in His Front Yard*, Portland, Ore. 1949

____________, *Henry Harmon Spalding, Pioneer of Old Oregon*, Caldwell, Idaho, 1936

____________, *Elkanah and Mary Walker, Pioneers Among the Spokanes*, Caldwell, Idaho, 1940

Durham, N.W., *History of the City of Spokane and Spokane County, Wash.* 3 vols. Spokane., WA, 1912

Edwards, Rev. Jonathan, *History of Spokane County*, 2 vols. W.H. Lever, Pub. 1900

Elsensohn, Sister M. Alfreda, *Pioneer Days in Idaho County*, (O.S.B.) Caldwell, Ida. 1947

Evans, Pauline, *The Story and Life of Jane Timothy Silcott, A Nez Perce Indian Princess*, (Typescript, author's possession)

From Wheat to Flour, Wheat Flour Institute, Chicago, Ill. 1965

Fuller, George, *History of the Pacific Northwest*, 4 vols. Spokane, 1928

Garfield County, Court House Records, 1877, Garfield, WA

Gilbert, Mr., *History of Whitman County, State of Washington*, W.H. Lever, Pub. 1901

Glover, James, Manuscript Reminiscences, MS in Eastern Washington State Museum, Spokane, WA (n.d.)

Glover, Richard, *David Thompson's Narrative, 1784-1812*, Champlain Society, Toronto, 1962

Haines, Francis, *Red Eagles of the Northwest, the Story of Chief Joseph and His People*, Portland, Ore. 1939

Jessett, Thomas E., *Spokan Garry*, Minneapolis, 1960

Kuykendall, Judge Elgin V., *Historic Sketches of Asotin County*, Clarkston, WA. 1954

Vanished Gristmills

Bibliography

Lyman, W.D., *History of Walla Walla Co.*, Vol. 1-2, W.H. Lever, Pub. 1901

__________, *History of Yakima Valley*, S.J. Clarke Pub. Co., 1919, Vol. I

Orchard, Vance, *Waitsburg, One of a Kind*, Waitsburg, WA. 1976

Polk's Directory, Nez Perce, Asotin County, WA, 1916-17

Pomeroy Commercial Club, *Garfield County, Soil, Climate, Resources and Opportunities*, Pomeroy, WA. (n.d.)

Puget Sound Business Directory and Guide to Washington Territory, 1872, Table of Gristmills

Rawson, Marion, *Little Old Mills*, E.P. Dutton and Co. N.Y., 1935

Sanford, Gregory, The Pacific Northwesterner, Chief Timothy, Vol. 12, no. 2, pp. 17-24

Steele, B.F. et al, *History of Stevens, Ferry, Okanogan and Chelan Counties, State of Washington*, Spokane, 1904

Story of Flour, Pillsbury Flour Co., Minn., 1922

Told by the Pioneers, W.P.A. Project, Sponsored by Federal Govt. No. 5841, Olympia, WA., 1938

Tucker, Gerald, *History of North Blue Mountains*, U.S. Dept. Agriculture, 1940

Winther, O.O., *The Great Northwest*, a History, New York, 1947

Woodbury, George, *John Goff's Mill*, Horton Pub. Co. N.Y. 1948

Work, John, *The Journal of John Work*, ed. W.S. Lewis and P.C. Phillips, Early Western Journals, 1, Cleveland, 1923

Zimiles, Martha and Murray, *Early American Mills*, New York, 1973

Florence E. Sherfey

Acknowledgements

Whitman Mission National Historic Site, Larry J. Waldron, Chief Park Interpreter, Walla Walla, Washington.

Smithsonian Institution, Division of Mechanical & Civil Engineering, Carlene E. Stephens, Museum Specialist, Robert Vogel, Curator, Wash. D.C.

Penrose Memorial Library, Whitman College, Lawrence L. Dodd, Curator of Manuscripts and Special Collections, Walla Walla, Wash.

Departmant of Conservation and Economic Development, Division of Parks, Richmond, Virginia, Ben H. Bolen, Commissioner

Cheney Cowles Library, Eastern Washington Historical Society, Mrs. William Kelly, Librarian, Spokane, Washington

Whitman County Historical Society, Roy M. Chatters, Pullman, Washington

Washington State Library, Nancy Pryor, Washington Room Librarian, Olympia, Wash.

Garfield County Historical Society, Mrs. Marion Beale, President, Pomeroy, Wash.

Whitworth College Library, Spokane, Wash.

Spokane Public LIbrary, Thomas E. Verd, Reference Librarian, Spokane, Wn.

Holland Library, Washington State University, John Guido, Archives Librarian, Pullman, Washington

Kennedy Library, Jay Rea, Archives Librarian, Eastern Washington University, Cheney Wn.

Washington Water Power Company Library, Spokane Wash.

Historical Society, Cheney, Wash. Mrs. A.B. Cutting, Curator

Perkins House Museum, Colfax, Washington

Stevens County Historical Society

Bureau of Land Management, Portland, Oregon

General Services Administration, Nat'l Archives and Records, Washington, D.C.

Paramount Photo-Lab, Spokane Washington

Acknowledgements

Individual Acknowledgements:

Mrs. William Houser, Mrs. Raleigh Hampton, Mrs. Marion Beale, Mr. Robert Weatherly, Mr. Robert Beale, Mr. and Mrs. Carl Silver, Mrs. Sydney Campbell, Mr. Keith Williams, Mr. Roger Oscarson, Mr. Joseph Barron, Mr. Ole Dybdall, Mr. Cecil Hutchens, Mr. Bennett Carter, Mrs.Wyoma Duckworth, Mrs. Edith Hill, Mr. Art Hunsperger, Mr. J. Weber Leid, Mrs. A.B. Cutting, Mr. John F. Carter, Mr. Ralph Shawn, Mr. Roy M. Chatters, Mr. J.B. West, Mrs. Mary Flood, Mrs. Delores Fels, Mr. Mickey Spitzenberger, Mr. Henry Smith, Mrs. Bettie Chase, Mr. Lawrence L. Dodd, Mr. Gerald Robeson, Mr. Jack Robeson, Mr. Glen Adams, Mr. Donald Ledgerwood, Mr. Harold Graham, Mr. and Mrs. George T. Dart, Mrs. Clifford Butts (Anna Baldwin), Dr. Ewart Baldwin, Mr. Cliff Wilson, Mrs. William Kelley, Mr. Frank M. Martin, Mr. Earl Victor, Dr. R.K. Allen, Mr. Nick J. Manring, Edward J. Kowrach, S. J.

Periodicals:

Walla Walla Union Bulletin, Walla Walla, Wn.
The Times, Waitsburg, Wash.
Spokesman Review, Spokane, Wn.
Spokane Daily Chronicle, Spokane, Wn.
Colfax Gazette, 1877 Reprint, Colfax, Wn.
Palouse Republic, Palouse, Wn.
Lewiston Tribune, Lewiston, Idaho
East Washingtonian, Pomeroy, Wash.
Pacific Northwest Forum, Spring, 1976
N. Kittitas Co. Tribune, Cle Elum

Index

Index

Index

Index

Index

Index

Index

Index

Index

Index

Index

Index

Index

Index

Index